About the Author

Anthony Winterbourne has written widely on Kant and epistemology. His articles have appeared in *Studies in History and Philosophy of Science*, *Historia Mathematica*, and *Kant-Studien*, among other journals, and he has published three further books. Born and educated in London, Dr. Winterbourne holds degrees in philosophy from the University of Bristol. He now lives in Norfolk, and devotes much of his time to writing and research.

The cover shows a drawing by Saul Steinberg which appeared first as a frontispiece to Jonathan Bennett's book *Kant's Analytic* (Cambridge University Press, 1966) and is Steinberg's commentary (as Bennett so aptly expressed it) "on the uneasy relationship between the a priori and the empirical." As such it is particularly appropriate here.

THE IDEAL AND THE REAL

Kant's Theory of Space, Time and Mathematical Construction

A T WINTERBOURNE

Published 2007 by arima publishing

www.arimapublishing.com

ISBN 978-1-84549-198-7

© A T Winterbourne 2007

All rights reserved

This book is copyright. Subject to statutory exception and to provisions of relevant collective licensing agreements, no part of this publication may be reproduced, stored in a retrieval system, or transmitted in any form or by any means, without the prior written permission of the author.

Printed and bound in the United Kingdom

Typeset in Garamond 11/14

This book is sold subject to the conditions that it shall not, by way of trade or otherwise, be lent, re-sold, hired out, or otherwise circulated without the publisher's prior consent in any form of binding or cover other than that which it is published and without a similar condition including this condition being imposed on the subsequent purchaser.

Abramis is an imprint of arima publishing

arima publishing
ASK House, Northgate Avenue
Bury St Edmunds, Suffolk IP32 6BB
t: (+44) 01284 700321

www.arimapublishing.com

CONTENTS

Preface   vii

CHAPTER ONE
Prologue: Newton and Leibniz

1.1. Newton on Space, Time and Metaphysics   1
1.2. Leibniz: The Ideal and the Real   27

CHAPTER TWO
Kant's Theory of Space and Time

2.1. Introduction   53
2.2. Concepts and Definitions   55
2.3. Kant's Anti-logicist programme   62
2.4. Transcendental Aesthetic   67
2.5. Construction and Schematism   79
2.6. Spaces and Geometries   97
2.7. Incongruent Counterparts and the Intuitive Nature of Space   104
2.8. Infinity: Reason and Experience   118
2.9. Transcendental Idealism   134

CHAPTER THREE
Acts, Intuitions and Constructions

3.1. Introduction   145
3.2. Concepts, Intuitions and the Schematism   147
3.3. Kant's Constructivism   158
3.4. Incongruity and Constructions   163
3.5. Indirect Proof   171

Bibliography   175
Index of Names   181

Preface to this Edition

This book was first published some years ago in the Nijhoff International Philosophy series by Kluwer Academic. In that format it was prohibitively expensive and has thus remained for some time beyond the pockets of its target readership of undergraduate and graduate students in philosophy. I have taken the decision to have the book re-printed in this affordable paperback version in order to make it freely available again to that audience. This is not in any sense a major revision of the original text: I have reluctantly concluded that such a task was impracticable for me at the present time. Moreover, it seemed to me that such a revision would probably have resulted in my producing a quite different book, and in the process losing this essay's character as being primarily introductory. Nonetheless, the text as first published has been completely re-cast so as to make for greater ease of reading. In this new form I believe that the work is now much more accessible, and that it will prove to be a useful and even occasionally valuable resource for students of Kant.

A. T. Winterbourne

Revised Preface to First Edition

The present work is predicated on the assumption that Kant's claims about space and time remain of intrinsic philosophical importance, and that those claims also assist in understanding what may eventually come to be regarded as Kant's most original contribution to philosophy, outside of the founding of the idea of a critique as such. I have in mind the distinction in Kant between the respective methods of philosophy and mathematics; and the further claim as to the nature of mathematical reasoning as having to do with construction. The fact that these claims retain an interest for philosophers and philosophically-minded mathematicians can only serve to remind us of the depth and originality of Kant's original thesis, given his supposed commitment to a view of mathematics long-since superseded.

Originality is an irreducibly conservative business. What we owe to Kant—even Kant at his best—is not fully explicable independently of what was bequeathed to him by the rationalist and empiricist debates of the previous few generations. It must therefore be true that we may not fully grasp Kant's achievement without having the occasional glance over his shoulder at his immediate predecessors. In this book I have assumed that there is actually very little point in trying to convey what Kant said about space and time without providing the essential background debate. Not only do Newton and Leibniz provide the best examples of Kant's problems, they also exemplify the shifting relationship between what is "real" and what is merely "ideal" which is the point of the title of this book.

There are of course many ways of coming to terms with the relationship between the ideal and the real, and the framework chosen in the following pages makes no claims to great generality or uniqueness. Nonetheless, the problem of space and time does seem to lend itself to being the focus of such a discussion. Spatial and temporal experience is, after all, simultaneously completely familiar, and deeply puzzling. And this very familiarity can often seem denied by mathematical idealisations of that experience which nonetheless still employ familiar terms. The "solution" to the problem of space and time offered by Kant—that they are "intuitions", not concepts—is, like almost everything else in the Critical Philosophy, an attempt to supersede the debate between rationalism and empiricism. In presenting this solution, Kant also offers to make intelligible the connection between the real and the ideal from within mathematics.

In this essay, I have tried to show the continuity existing between the solution to the problem in general terms, and the solution offered for the special case of mathematical reasoning. So far as space and time are concerned, Kant's perception of the problems involved can most easily be grasped in relation to the debate between Newton and Leibniz. In what follows I have tried to give some sense of the importance not only of the polemics, but also of the respective positions taken. In Newton's case, in spite of the "theological" or metaphysical tone of much of his speculations on this subject, I suggest below that his specific and notorious arguments for absolute space in particular, can in fact be read less abstractly, and more "analogically" than is usually the case. For all that, Newton remains committed to a view of space and time which seems a long way from agreeing with his "positivist" line taken on other philosophical issues. There is therefore something of an irony in the fact that while Newton's "empiricism" deserts him on this point, Leibniz's rationalism leads him to explain space and time in phenomenal terms. However, I will in fact be arguing that the traditional characterisation of Leibnizian space and time as phenomena is misleading. For Leibniz, the problem of the relationship between the real and the ideal will be explained by finding room for space and time locutions on three different metaphysical/ontological levels.

The transformation of these ideas effected by Kant will inevitably require not only a re-casting of the traditional problems, but will further require an important shift in nomenclature. Throughout this essay I have tried to do justice to the subtlety of Kant's ideas in a fairly short space. It is my hope that this relative brevity does not lead to Kant's position on these issues being grossly distorted.

Kant's unique solutions to the problems he inherited are the result of two distinct strategic innovations: first, he will re-order the relationship between the phenomenal and the noumenal, effectively reversing the epistemological polarity as expressed in Leibniz; and second, through his insistence on an irreducible difference between the methods of philosophy and mathematics respectively, he will provide an intra-theoretical contrast between the ideal and the real within the province of the latter discipline. Since these two strategies are profoundly interwoven in Kant's writings, any conclusions or speculations made on the evidence of one set of considerations will have an inevitable effect on the other set. As a consequence, even though the "ideal/real" locution has been employed below as the most appropriate for talking about Kant's philosophy of mathematics, it is nonetheless sufficiently general to capture the spirit of the critical philosophy of space and time as such.

I have suggested in this book that Kant's solution to what I have called the "application problem" in the general case of relating pure concepts to intuitions, can profitably be approached through the special case of relating ideal mathematical objects to real mathematical constructions. In both instances, what Kant requires is the mediating function of "schemata." My purpose in those sections dealing directly with this issue is to remove some of the artificiality and obscurity from this notion. I make no extravagant Heideggerian claims to having identified the key to Kant's thought as a whole: in the brief compass I have taken, I hope modestly to have brought together a number of crucial Kantian arguments and suggested that an authentic Kantian theme can be found to connect them.

For the reader's convenience, all purely textual references are made directly in the body of the work to those books listed in the Bibliography. References to Kant's works are given after each quoted passage, using the following convention: references to the *Critique of Pure Reason* are made using the 'A' and 'B' numbers directly; other abbreviations are used as follows:

P.C.—*Pre-Critical Writings.*
Prol.—*Prolegomena.*
C.Pr.R.—*Critique of Practical Reason.*
C.J.—*Critique of Judgement.*
Corr.—*Philosophical Correspondence.*

A.T.W.

Chapter One

PROLOGUE: NEWTON AND LEIBNIZ

1.1. Newton on Space, Time and Metaphysics

Newton's writings on metaphysics, especially in so far as these impinge on his scientific concerns, are widely scattered, and are often little more than assertions rather than philosophical arguments. It is therefore of the highest importance that we have available in the Clarke-Leibniz correspondence a sustained and detailed argument which covers all the outstanding metaphysical problems raised by Newton in his *Principia* and elsewhere. There is ample evidence that Clarke was acting not simply as an interpreter of his master's views, but was actually writing under instructions from Newton, although Clarke does sometimes express views which Newton would probably have rejected. Certainly Leibniz was in no doubt as to whom his real adversary was. [In H.G.Alexander, p.189]

Newtonian physics, according to many commentators, must stand or fall with the concepts of absolute space and time. In so far as Newtonian dynamics rests upon the laws of motion, and the laws of motion in turn presuppose "immovable space", such an interpretation would seem to have much plausibility. The laws of motion demand an infinite, homogeneous space, independent of the bodies in motion in that space. In spite of attacks made on the notions of absolute space and time, the system of dynamics survived, quite simply because it succeeded in explaining and predicting phenomena where the alternatives did not. It was not until Einstein's rejection of the concept of absolute simultaneity—which presupposes absolute time—that a new physics arose. Jammer has remarked that all of Newton's achievements and discoveries in the realm of physics are subordinate to the philosophical conception of absolute space. [Jammer, p.114]

However, we must be careful to distinguish the Newtonian versions of these concepts into those required by his dynamics, and those employed—as hypostatisations—in theological and metaphysical contexts. Newton was undoubtedly influenced in some sense by Henry More's theory of space and time:

> This infinite extent will seem to be something divine. It cannot be nothing as it has so many splendid attributes, such as the following, which are ascribed by metaphysicians to the Supreme Being. One, Simple, Motionless, Eternal, Complete, Independent, Pre-Existent, Subsisting of Itself, Incorruptible, Necessary, Immense, Uncreate, Not bounded, Incomprehensible, Omnipresent, Incorporeal, All Pervading and Embracing, Essential Being. [In Vasiliev, p.35].

Newton also regarded certain dynamical arguments as decisive, and his non-scientific convictions exercised a profound influence on the final form his ideas were given. Whether we are to regard the purely scientific aspects as more important for Newton than the theological is not a question that admits of an obvious answer, especially given his preoccupation with both theological and alchemical issues. [Dobbs, p.32]. Did the absoluteness of space lead to its divinisation, or vice versa? I think it is reasonable to assume that while Newton accepted some version of space and time as the *sensorium Dei* before he discovered what he regarded as crucial arguments from dynamics, the latter may have strengthened his pre-scientific (and non-scientific) convictions in a radical way.

In an early and very important manuscript, *De Gravitatione et Aequipondo Fluidorum,* Newton discusses the nature of space and time in more detail than anywhere outside *Principia,* not counting, for these purposes, Clarke's letters to Leibniz. This manuscript contains important clues to understanding his later theories. Space, Newton writes, is not a substance because it is not absolute in itself, but it is rather "an emanent effect of God, or a disposition of all Being." It is not to be regarded as an attribute, and yet cannot be said to be nothing at all: it is rather "...something, and approaches more nearly to the nature of a substance." [in Hall, p.132]. Newton wishes to show not only that space *is* something, but also what it is. For Newton, space must be actually infinite. He does not seem to have been unduly troubled by the concept of infinity (unlike Leibniz), and does not discuss it often, though it is often presupposed, nowhere more obviously than in relation to space and time. He recognises a distinction between imagination and conception—a distinction of faculties. "We can imagine a greater extension, and then a greater one but we understand that there exists a greater extension than any we can imagine." [in Hall, p.134]. Newton rejects the view that our idea of the infinite can be derived from the mere negation of the finite, and in fact argues for the converse of this. By negating all limits the conception becomes positive:

> 'End', *(finis)* is a word negative to sense, and thus 'infinity' (not-end) as it is the negation of a negation, (that is, of ends) will be a word positive in the highest degree with respect to our perception, though it seems grammatically negative. [in Hall, p.135]

Indefinite, according to Newton, is not a term which can be applied to things as such, but only to our limited conception, i.e., it means "not yet determined by us": it therefore takes on the character of an epistemological rather than an ontological category. The infinitude of space follows necessarily from its homogeneity: if space is homogeneous, then all points in it are exactly similar—none can have any privileged status; if space was finite then there would be some points in it—those "at the limits"—which differed fundamentally from some others. The acceptance of the homogeneous nature of space is a complex assumption involving mathematical rather than phenomenological considerations, and forms the cornerstone of the classical conception of nature within which Newton has his roots.

Absolute space, in its own nature, without relation to anything external, remains always similar and immovable. Relative space is some movable dimension or measure of the absolute spaces; which our senses determine by its position to bodies; and which is vulgarly taken for immovable space. [in H.G.Alexander, p.152.][1]

Thus Newton does not reject the notion that space is relative; what he rejects is the notion that it is only relative. Relative space is phenomenal, but this space of the world of perception is not ultimate. It is also the space of the science of kinematics, in so far as this is merely the "movable dimension" or measure of the absolute spaces. We see here the abandonment by Newton of the positivistic side of his philosophical temperament, with this assertion of the necessity for abstracting from sense-experience. The Laws of Motion, the Definitions, and Scholium, are the fundamental presuppositions on which the edifice of the *Principia* will be constructed. At this stage Newton is setting out the idealised concepts of space, time, and matter in motion, without which the new science of dynamics would be impossible. These are idealised concepts, yet they have their roots in sense-experience, of which they are reconstructions or "re-descriptions". Whereas Kant will say that space and time must be presuppositional forms in order for experience as such to be possible, Newton is suggesting here something to the effect that only if space, time, and motion are like this, is the *Principia* as a whole possible; since *Principia* is true it is, a fortiori, possible; thus space, time, and motion, as given by Newton, are supposedly vindicated. There is a suggestion of the acceptance by Newton of the possibility that space and time are presuppositional forms in the sense in which Kant would take this, though this is not in evidence in the early manuscript *De Gravitatione* as an openly argued thesis. There he says that neither space nor duration are completely independent of Being, for "...when any being is postulated,

---

[1] For convenience, all references to the famous Scholium will be to the translation offered along with the Clarke-Leibniz correspondence in H.G.Alexander, Manchester University Press, 1970.

space is postulated. And the same may be asserted of duration." (It is interesting to compare this with Kant's denials of the possibility of "thinking away" space and time from their sensuous content.)[2] Newton reconciles this with his notion of space and time as independent of bodies by means of the theological idea of an omnipresent and eternal God.

So for Newton, absolute space and time remain always similar and immovable. That space must remain always similar and immovable follows from its homogeneity and infinitude; only the finite is movable, and motion is intelligible only with reference to space, whether absolute or relational. Absolute space could not move with reference to itself, and if it did "move", it would not be absolute space at all.

For Newton, "body", and "place", are correlative. Place is a part of space that a body fills, while "body" is that which fills a place. But this "part" of space is not the same as the body's situation, nor the external surface of the body. Such a notion of body is itself an idealisation from perceived bodies: the properties that define this idealised concept are extension, mobility, and impenetrability. By defining a body as that which fills a place, and place as a part of space, Newton has logically separated bodies from space. Since extension is one of the fundamental properties of matter, the property of occupying a spatial region must belong to bodies fundamentally. [In H.G.Alexander, p.152].

Absolute time is an even more difficult concept than space:

> Absolute, true and mathematical time of itself, and from its own nature, flows equably without relation to anything external, and by another name is called duration: relative, apparent and common time, is some sensible and external (whether accurate or unequable) measure of duration by the means of motion, which is commonly used instead of true time; such as an hour, a day, a month, a year. [in Hall, p.128].

It is very difficult to know how this is to be understood. The idea of absolute time "flowing" in an otherwise empty universe is unintelligible in

---

[2] This point will be discussed in more detail in 2.4., below.

a way in which the "container" theory of space is not. What sense can be given to the idea of time not flowing? Time seems to be more obviously connected with change, process, or consciousness, and the separation of time from events or objects is an idea which is bound to seem mysterious if not actually meaningless.

But for the present I want to return to absolute space and motion. The concept of motion is Newton's central problem in natural philosophy, and is the theme of *Principia* as a whole. The answers given there depend upon the intelligibility of the notions of absolute space and time. Conversely, the existence and nature of the latter are validated by the arguments from motions and, as we shall see, forces. In Newtonian dynamics it is only changes of motion that require special explanations; motion—uniform motion—and rest, are ontologically similar. Motion becomes a genuine state, and its opposition to rest a matter of correlation.

To state the distinction between absolute and relative motion is easy: to fully understand the distinction is not. Absolute motion is the translation of a body from one absolute place to another. A body which is really at rest, is at rest with respect to absolute space. To talk of "relative translation" is actually misleading here; a body a may change its relative situation or position with respect to a second body b when b orbits a. But it is misleading to think of this as a translation of a from one relative place to another. In *De Gravitatione* Newton says that "...physical and absolute motion is to be defined from other considerations than translation, such translation being designated as merely external." [Tamny, p.57 note.]

I now want to examine Newton's "proofs" of absolute space. I shall argue that Newton's argument has been misrepresented, and that what the Scholium to Definition VIII of *Principia* says should be rendered rather more literally than has been the case. What I shall be offering is a less "metaphysical" reading of these famous passages. That there should indeed be a more straightforwardly epistemological rendering of these arguments has also been suggested by Martin Tamny, who writes:

CHAPTER ONE – PROLOGUE: NEWTON AND LEIBNIZ

"...Whereas all previous arguments drawn from Newton concerning this claim have been purely metaphysical and theological, we have seen that there may well have been epistemological considerations as well."[3] It is my intention to take up this suggestion. I will go further and argue that Newton's original argument has been idealised, and that the logic of the globes thought-experiment has been changed by commentators. When this changed logic is attributed to Newton, important aspects of his reasoning have been overlooked, and a more literal reading of these discussions obscured.

It is usually accepted that since Newtonian mechanics requires an absolute reference frame, Newton must defend that conception from within the framework of natural philosophy or be compelled to present this reference frame as an epistemologically unjustified, but somehow metaphysically necessary, presupposition.[4] The orthodox view is that Newton uses the arguments of the spinning bucket, and the cord-connected spheres, as evidence that "empirically" verifies the existence of absolute space. In fact it is not explicitly stated by Newton that these arguments prove the existence of absolute space: this conclusion is nonetheless said to follow directly from the fact that they demonstrate absolute motion. The orthodox reading of Newton's argument is well illustrated by Max Jammer, who cites this passage from *Principia*: "The causes by which true and relative motions are distinguished one from another, are the forces impressed upon bodies to generate motion." [In H.G.Alexander, p.156]. (For clarity, I will refer to this as N1, i.e., "Newton's First Thesis".) What Jammer calls Newton's first argument is based on the idea that only forces generate real motion. According to Jammer, this argument is constituted by three significant passages. The first is the one I have called "N1"; the second is this:

---

[3] That is, that absolute space may at best be regarded as a 'useful fiction'. See also A. d'Abro, especially p.106ff and 412ff. See also John Losee, p.84ff.
[4] This view that the space-time Scholium was not intended as a proof of absolute space has also been argued by Laymon.

The effects which distinguish absolute from relative motion are the forces of receding from the axis of circular motion. For there are no such forces in a circular motion purely relative, but in a true and absolute circular motion they are greater or less according to the quantity of the motion. [in H.G.Alexander, p.157.]

(I shall refer to this passage as "N2".) The third passage cited by Jammer as constitutive of Newton's first argument—"the argument from forces"—is this:

It is indeed a matter of great difficulty to discover, and effectually to distinguish, the true motions of particular bodies from the apparent; because the parts of that immovable space, in which those motions are performed, do by no means come under the observations of our senses. Yet the thing is not altogether desperate; for we have some arguments to guide us, partly from the apparent motions, which are the differences of the true motions; partly from the forces, which are the causes and effects of the true motions. [In H.G.Alexander, p.159.]

(For reasons which will become apparent later, I shall refer to this passage as "N4".) Newton states quite clearly here that the problem is to distinguish true and relative motion, given the existence of absolute space. The function of the thought-experiments is to show that any relational mechanics leaves out something important in the description of certain kinds of motion, giving indirect support to the idea of an absolute reference frame.

So much for what Jammer calls the first argument. The second argument for the existence of absolute motion is "...that which proceeds from the effects that such motion produces, in particular the appearance of centrifugal forces. So we have Newton's famous pail experiment." After discussing this Jammer insists that "...the same inaccessibility to physical verification characterizes all the other attempts to enforce [Newton's] argument, as for example, his experiment with the two cord-connected spheres." [Jammer, p.106]. Newton's final argument, based on the distinction between absolute, and relative or apparent motion is not, in Jammer's opinion, developed further. On Jammer's interpretation then, the logical (or epistemological) status of both thought-experiments is the same. Both are attempts to move from the existence of forces to the existence of absolute space, via absolute motion. It is this interpretation that I intend to question. There is in addition, as already noted, a tendency to change the logic of the globes experiment in particular, and attribute this changed logic to Newton. For example, Lawrence Sklar describes the globes experiment in what has become the accepted fashion, thus idealising Newton's original case. The orthodox manner of considering the argument is based on what are taken as the initial conditions given in the Scholium: that is, something like "...consider two globes, connected by a cord, in an otherwise empty universe." [Sklar, (1974), p.183.] In fact, the implications of considering a universe empty except for the globes, concerns Newton's extensions of the argument: this particular case, as described by Sklar (and others) is what issues from the argument—it is not an initial condition of it.

To support my thesis that the orthodox reading of the relevant Scholium passage is, if not strictly-speaking wrong, then at least not Newton's view, let us return to Jammer's discussion. What I have referred to above as N1, N2, and N4, are run together by Jammer as representative elements of a single argument. Perhaps the most often cited passage of the Scholium, referring to the introduction of the two thought-experiments, in fact introduces only the argument of the globes. Most commentators present this as though the passage beginning "...[i]t is indeed a matter of great difficulty to discover...the true motions of particular bodies from the apparent..."; and which includes the significant assertion, "...yet the thing is not altogether desperate...", preceded the bucket experiment. But it does not. It comes after that discussion, and immediately before the introduction of the globes thought-experiment. I hope to show that this is an important and neglected aspect of Newton's reasoning, and that the Scholium passages might be taken rather more literally than is usually the case.

Now I agree with Jammer's suggestion that the bucket experiment focusses attention on the special role of certain forces and their observable effects. According to Newton, these can only be explained by postulating an absolute reference frame. I think it is clear that the notion of force is really what the Scholium passages are about, and it is this notion—as a fundamental reality on the phenomenal level—which most embarrasses Leibniz in the correspondence with Clarke. Once Leibniz has admitted that it is the attribution to a body of force that determines whether it is really in motion, he has a made a major concession to Newton's position. Leibniz's problems are intimately related to the different levels on which force and motions are realities in his metaphysics. On the phenomenal level Leibniz adheres to a mechanical explanation: on the other hand, force is something that a body possesses as part of its intrinsic metaphysical nature. (These issues will be discussed in more detail in 1.2., below.)

Newton thinks that explaining the role of certain forces in relation to an absolute reference frame is accomplished by pointing out that in true circular motion certain forces are greater or less, and are measurable. What I have called N1 is part of an argument from dynamics designed to show that true motion cannot be explained relationally. From the relational viewpoint, the translation of bodies can only be seen as a kinematic change which means, for Newton, only by means of relative or sensible measures. Since he is concerned to show that the confounding of these measures with the "real" measures generates "vulgar prejudices" concerning space and time, his task is to establish that although certain motions are kinematically equivalent, they are nonetheless dynamically distinguishable. This distinguishability is manifested by the measurable variations of certain forces. Hence the bucket experiment. And Newton expresses no doubts about this example or its results: "There are no such forces in a circular motion purely relative, but in a true and absolute circular motion, they are greater or less, according to the quantity of the motion." [In H.G.Alexander, p.157.] A dynamic explanation must be

CHAPTER ONE – PROLOGUE: NEWTON AND LEIBNIZ

relative to a frame of reference other than the "ambient bodies."[5] Newton's argument is always from the existence of observable forces, or their effects, to absolute motion, and thence to absolute space. Although the latter cannot be observed, it can be inferred as a way of making these forces and effects intelligible. Newton regarded it as the simplest hypothesis to explain the observed facts of mechanics. Absolute motion is motion which cannot be regarded as relative to anything observable such as the ambient bodies, or matter in general.

Newton advances his argument by repeating what he regards as the error of confusing sensible measures with real entities:

> Wherefore relative quantities are not the quantities themselves...but those sensible measures of them (either accurate or inaccurate) which are commonly used instead of the measured quantities themselves. And if the meaning of words is to be determined by their use, then by the names time, space, place and motion, their (sensible) measures are to be properly understood; and the expression will be unusual and purely mathematical, if the measured quantities themselves are meant. On this account, those violate the accuracy of language, which ought to be kept precise, who interpret these words for the measured quantities. Nor do those less defile the purity of mathematical and philosophical truths, who confound real quantities with their relations and sensible measures. [In H.G.Alexander, p.158-9.]

It is after this passage (which I shall call N3) that Newton asserts that it is difficult to discover and distinguish true motion from apparent, yet reassures his readers that "the thing is not altogether desperate". Now if we accept the conventional reading, it would clearly be an inappropriate moment for Newton to say this: that is, *after* consideration of the bucket experiment which, on the orthodox reading, establishes all Newton's purposes in the Scholium arguments. The most plausible interpretation seems to be that Newton has tried to establish that a full explanation of certain phenomena must include reference to forces differentiating real

---

[5] Laymon suggests that a proper reading of 'ambient' in this connection is one which expresses Newton's anti-Cartesianism. See Laymon, p.404.

and apparent motion. And "...the thing is not altogether desperate; for we have some arguments to guide us, partly from the apparent motions, which are the difference of the true motions; partly from the forces, which are the causes and effects of the true motions." It is possible to guess at what Newton means by the "differences" of the true motions, by looking at what MacLaurin made of this in his exposition of Newton's thought. MacLaurin writes that: "In general, the actions of bodies upon each other depend not upon their absolute but relative motion; which is the difference of their absolute motions when they have the same direction, but their sum when they are moved in opposite directions." [MacLaurin, II, p.128.] This is what Newton seems to have in mind in the following: "But if the earth also moves, the true and absolute motion of the body will arise, partly from the true motion of the earth, in immovable space, partly from the relative motion [of the ship] on the earth." [In H.G.Alexander, p.153.] Now why should Newton, at this point, say that there are some arguments to guide us if, as is commonly supposed, the bucket experiment constitutes the argument? Immediately following the last quoted sentence Newton introduces the globes thought-experiment. So does it not seem plausible that Newton regards N4 as offering two kinds of argument, which though obviously related, are not intended to be making precisely the same point? Suppose we assume that the bucket experiment argues for the special role of certain forces: we might then consider the possibility that the globes experiment is Newton's argument "...from the apparent motions, which are the differences of the true motions."[6]

---

[6] I shall pursue this line of argument in despite the fact that Newton's words here reverse the order of the experiments. It is much harder to make sense of the Scholium discussion if the distinction I am making here is not followed. If Newton does, as I believe, intend just such a distinction, this seems the likeliest place to identify the beginning of the argument. The distinction has, it seems to me, been confused, partly by running together passages N1, N2, and N4, and using them as justification for the bucket experiment, without the mediation of N3.

CHAPTER ONE – PROLOGUE: NEWTON AND LEIBNIZ

The details of the bucket experiment are as follows. It consists in suspending a bucket of water by a rope which is then twisted. A force is applied to the bucket in the direction of unwinding. Newton writes:

> While the cord is untwisting itself, the vessel continues for some time in this motion; the surface of the water will at first be plain, and, as before the vessel began to move; but after that, the vessel, by gradually communicating its motion to the water, will make it begin sensibly to revolve, and recede by little and little from the middle, and ascend to the sides of the vessel, forming itself into a concave figure...and the swifter the motion becomes, the higher will the water rise, till at last, performing its revolutions in the same times with the vessel, it becomes relatively at rest in it. [In H.G.Alexander, p.157.]

Newton's central point is that the resulting deformation of the water surface indicates the existence of forces acting. Of course this is not, in itself, surprising, since the experiment specifies the application of an external force to the bucket. The essential point is that certain kinds of forces, when applied to systems free to rotate, or to systems already rotating, give rise to centrifugal forces as their effects. The second law of motion associates accelerations with forces. With respect to what, then, is the water accelerating? Newton insists that it cannot be the bucket, since the water surface is successively plane and concave when there is relative acceleration, and since the surface may be plane or concave when there is no relative acceleration. He concludes, not that the acceleration must be with respect to absolute space—this term nowhere appears in the relevant discussion in the Scholium—but that

> ...this endeavour (of receding from the axis of rotation) does not depend upon any translation of the water in respect of the ambient bodies, nor can true circular motion be defined by such translation. There is only one real circular motion of any one revolving body, corresponding to only one power of endeavouring to recede from its axis of motion, as its proper and adequate effect. [In H.G.Alexander, p.158.]

C.D.Broad suggested that since the term "absolute space" nowhere appears in the premises of Newton's argument it cannot, logically, appear in the conclusion. [Broad, (1923), p.100ff.] In fact the term occurs neither as premise nor conclusion. The argument is, as I have indicated, about force—both as a cause of certain kinds of motion, and as an effect of such motion. It must be added that the fact that Newton is dealing here with a deformable body greatly complicates the issue. He says that when the relative motion of the water in the vessel is greatest, it produces 'no endeavour' to recede from the axis. The "it" here refers to the water considered as a single body, and Newton is trying to show that "...there is only one power of endeavouring to recede from its axis of motion, as its proper and adequate effect." However, since the water is deformable, it is not strictly true that there was no endeavour to recede from the axis at the beginning of the experiment. The particles of water in contact with the sides of the vessel would have received an immediate centripetal force, and would have started to rotate. The vessel, says Newton, "gradually communicates" its motion to the water; the endeavour shows that the real circular motion is continually increasing, and is at rest relatively to the bucket only when the whole body of water partakes of the motion, and thus acquires its greatest quantity. The concavity of the surface is a function of the combined circular motions of the particles constituting it. The true circular motion that Newton is trying to demonstrate is manifest only when the centripetal force has been communicated to the whole body of water. This takes time: when the circular motion of the water—considered as one body—is at a maximum, the water is at rest with respect to the sides of the vessel.

I come now to the globes thought-experiment. I want to suggest three ways to analyse this example. The first is the modern interpretation; the second is more firmly based on Newton, and concerns what might be called "Clarke's embellishment"; and the third is the argument I take it that Newton actually presented in the famous Scholium.

The modern version, which I shall only briefly outline, starts with the assumption of a universe empty except for the globes. In this case, the question of whether the globes rotate presents the problem almost from the outset as a matter of semantics. If one asks a relationist why we may not affirm that the globes rotate, his reply will be that the idea of circular—or any other kind—of motion is intelligible only when some other body is given as a reference standard. He may simply offer a challenge to the Newtonian to describe without question-begging the nature of rotation in an otherwise empty universe. The relationist presupposes that motion is intelligible only relationally; the Newtonian, in trying to make sense of the rotation of the globes, presupposes that the relational explanation is incomplete by neglecting the facts relating forces and tension. Of course, the Newtonian concedes the intelligibility of relational theories of motion: for him it is both meaningful and true—though not, of course, the whole truth. If the problem is seen like this, i.e., as a semantic dispute, it could be argued that Leibniz has the most consistent position, since the initial conditions of the experiment, viz., "consider an empty universe", would not pass unchallenged. However, this would be for metaphysical, rather than dynamic or "semantic" reasons: Leibniz would appeal to the metaphysical principles of "Perfection" and "Sufficient Reason."[7] This modern version of the argument has the merit of recognising that the problem can be reconstructed in terms of what it makes sense for us to affirm in certain well-understood cases of dynamics. It errs, I think, in attributing to Newton a too metaphysical cast of mind in this important argument.

The second version, though not found in the Scholium, is a recognisably Newtonian move. Consider the possibility that all objects apart from the globes are annihilated from the actual, non-empty universe. Does it make sense under these circumstances to talk about "rotation" and "direction of motion" of the globes? Newton—and Clarke, in the correspondence with Leibniz—have no hesitation in saying that it does. "[And] yet no way is shown to avoid this absurd consequence, that the parts of a circulating body would lose the

---

[7] Details of these metaphysical principles will not be given here. For extended discussions of them see, for example, Latta, p.21ff; and Martin, (1967), *passim*.

centrifugal force arising from their circular motion, if all the extrinsic matter around them were annihilated." [In H.G.Alexander, p.101.] There is no doubt I think that this way of stating the problem caused Leibniz some embarrassment. It would seem that he cannot agree with Newton without undermining his own concept of phenomenal motion—partly because of the uneasy traffic in his system between metaphysical and phenomenal levels of reality. (This complicates Leibniz's own theory of space and time in turn, as we shall see below.) If Leibniz denied that in the circumstances obtaining in Clarke's embellishment of Newton, there would be tension or rotation, he would commit himself to the view that an occurrence outside the globes could have an immediate effect on them; he would be committed, in other words, to action-at-a-distance of a quite radical kind—a thoroughly un-Leibnizian notion.

As I have said, the globes thought-experiment is not described by Newton in terms of which Sklar's account, for example, is typical. And this brings me to my third interpretation of the argument.[8]

[I]f two globes, kept at a given distance one from the other by means of a cord which connects them, were revolved about their common centre of gravity, we might, from the tension of the cord, discover the endeavour of the globes to recede from the axis of their motion, and from thence we might compute the quantity of their circular motions. [In H.G.Alexander, p.159.]

Evidently, at the point where Newton explains the purpose of the experiment and the results that might accrue, there is no suggestion that we should consider the globes in an otherwise empty universe. Newton is claiming only that by "testing" the cord for tension, we could compute

---

[8] Cf. also Ian Hacking: '[I]n *Principia* we are to imagine a universe with nothing in it but a bucket of water that starts to spin. This hypothesis can make no sense to a relativist, yet we know what would happen if it were true. Although the "spin" would not be visible, the water would gradually start to rise up the side of the bucket. Hence, even if there were nothing else in the world, there would be a difference between spin and rest, and so, said Newton, relativism is refuted.' (Hacking, p.249-50.) This is a distorted version of the argument of the Scholium; Newton nowhere suggests that we should consider the bucket in an empty universe.

the amount of circular motion. He argues that variations in the application of forces to alternate faces of the globes would either add to the quantity of circular motion or diminish it. Such computations would enable us to establish the quantity of the motion and its direction or determination, i.e., its vectorial value. It is at this point in the argument that Newton suggests for the first time that such computations would be possible "...even in an immense vacuum, when there was nothing external or sensible with which the globes could be compared." This extension of the conditions of the argument would, of course, not be intelligible to the relationist. The differences in the conditions given to us in various (Galilean) frames—exemplified in the argument from forces and accelerations—may be regarded either as intrinsic or as due to external influences. Newton—in his "postulate of isolation"—opts for the former. [See d'Abro, p.106ff.]

This is Newton's only direct reference to such a possible, non-actual universe. From this point he extends his reasoning by adding the fixed stars to this hypothetical universe. Even then we could not tell by means of the relative translations of the stars and the bodies which of them was really in motion: we have to acknowledge the kinematic equivalence of hypotheses. However, "...if we observed the cord and found that its tension was that very tension which the motions of the globes required, we might conclude the motions to be in the globes, and the bodies to be at rest." [In Alexander, p.159.] Now what can be meant by "...that very tension which the motions of the globes required..."? The tension must be such as to guarantee that the globes rotate about their common centre of gravity of course, so we might read this as "the tension required for orbital motion." But what Newton seems to be referring to here is his previous remark, where he thinks he has shown how one might come to the conclusion that variations in tension are caused by, and are directly proportional to, forces variously applied to the alternate faces of the globes. This is achieved by measuring the variations of tension observed in the actual, non-empty universe. Thus we could discover "...from the translation of the globes among the bodies", the determination of their motion. That is, once it is established that it is the globes that are in motion, not the stars, we can compute the direction of their motion with respect to the fixed stars considered as at rest.

For Newton, these remarks in the Scholium effectively embarrass any relationist of a Leibnizian persuasion who rejects action-at-a-distance. Newton has tried to show that the addition of forces on alternate faces of the globes has a direct and measurable effect on the cord's tension which is its "proper and adequate effect." It might be plausible to suggest that the connection between circular motion and centrifugal effects is sufficiently ambiguous such that we could not tell if rotating the fixed stars around the globes would result in the generation of such forces on the globes: but it is a good deal less plausible to say that by the addition of force on the body one affects the fixed stars. Indeed, Newton anticipated just this in *De Gravitatione* where he writes: "..who will imagine that the parts of the earth endeavour to recede from its centre on account of a force impressed only upon the heavens? Or is it not more agreeable to reason that when a force imparted to the heavens makes them recede from the centre of revolution thus caused, they are for that reason the sole bodies properly and absolutely moved." [In Hall and Hall, p.134]. It is not the existence of centrifugal effects as such that Newton believes refutes the relationist: in this case the bucket experiment would probably have been deemed sufficient. It is the law-like connection Newton thinks he has established between the variations in the amount of tension in the cord, and the variations in the applied forces. Having rejected kinematic relativity as explanations of these phenomena, the thesis must be developed mathematically and dynamically. What Newton has done thus far in the *Principia* is to indicate what is necessary to explain these causes and effects, and thus explain true and apparent motion. How this is to be done is the task of the work as a whole.

To summarise: the globes thought-experiment postulates the cord-connected spheres in the actual universe. In these circumstances we may refer to impressed forces, tension etc., without being open to the charge that anything we may say about such concepts is lacking in a clear sense.

It is important to keep in mind that any such assertions are based upon our knowledge of the globes themselves, considered in a normally-inhabited universe. From this we conclude that the impressed forces and variations of tension mutually imply one another in some law-like manner. It is this that Newton believes justifies the assertion that there would be such forces even in an otherwise empty universe. To deny this would be to deny not some metaphysical-theological thesis about absolute space, but the putatively established connection between forces and tension. Newton is extrapolating from the known behaviour of bodies in the actual universe, to the intelligibility of the idea of force in a possible world inhabited only by the globes. In other words, in such a universe, the science of dynamics would still apply—an assumption necessary for the laws of motion.[9] The analogical reasoning which leads to the acceptance of the intelligibility of forces in a universe empty except for the globes, helps us to grasp the logic of the laws of motion themselves. The assertion about the existence of forces and tension in the globes system holds also in the idealised case. The entire argument is "epistemological", although for Newton and (especially) his contemporaries, it was to have metaphysical—even theological—consequences. The two famous thought-experiments are therefore best treated as analogical arguments, and are less abstract and metaphysical than they are often taken to be. By treating both of Newton's examples in much the same way, commentators have missed interesting facets of the argumentation in the Scholium.

Most of the Scholium discussion is concerned with the existence and nature of centrifugal forces. It was an investigation of centripetal motion which led to Newton's formulation of the law of gravitational attraction. It is clear that the existence of centrifugal forces was regarded by Newton as a phenomenon requiring a special explanation. Indeed, the explanation

---

[9] The Laws of Motion can be taken as idealisations of the motions of observable and measurable bodies. Such observations lead to these laws, though the latter are not simply inductive generalisations from such observations.

was unique; such forces could only be explained by means of the idea of absolute space—the *sine qua non* of Newtonian physics.

What is the relation between centrifugal and centripetal force? It is common today to regard the former as purely fictitious—as a "delusion" arising from living in a rotating system and not allowing for its motion. According to Newton a centrifugal force is shown by an endeavour of a body to recede from the axis of motion. A centripetal force is that force by which bodies tend toward a certain point inwards. Gravity is thus a centripetal force.[10] Newton says that all bodies tend to recede from the centres of their orbits; were it not for the opposition of a contrary force which checks them, and retains them in their orbital motion, they would fly off in straight lines, with uniform motion. [See H.G.Alexander, p.148.] "Fly off" is somewhat misleading; the orbital motion requires a real force inwards, and without this force the body would continue its uniform rectilinear motion.

Before Newton, Huygens had suggested that centrifugal forces were the explanation for the phenomenon of gravity, a view shared by Descartes in his concept of vortices. As we have noted, Newton's concept of gravity came from a consideration of both centrifugal force and centripetal motion. However, for Newton what required explanation was not centrifugal force as such, but this and centripetal motion, which together enabled bodies to maintain their orbits. Rotation implies the existence of both forces, in so far as both forces are real. Circular motion, for Newton, must be conceived in terms of an equilibrium of forces. With an experiment of his own Huygens demonstrated that centrifugal force can produce centripetal motion—"an excellent example of a Cartesian vortex." Newton, of course, rejected the idea that gravity was an inherent property of bodies, and not content to offer hypotheses as to

---

[10] This raises the question whether Newton's original framing of the globes thought-experiment was ever adequate to his purposes. If gravity is a kind of centripetal force, the very idea of the globes having, from the outset, a 'common centre of gravity', already implies their rotation *and* the existence of centrifugal force. The two globes pull each other inwards, the pulls providing the needed centripetal forces.

20

CHAPTER ONE – PROLOGUE: NEWTON AND LEIBNIZ

its cause, restricted himself to a mathematical description of the phenomenon.

Why then does Newton feel it essential to offer causes for centrifugal effects—special causes, that is? Why not also limit himself here to pure description? The only answer can be that without the argument from centrifugal effects to absolute motion, the metaphysical basis of his system would be undermined. Richard Westfall has suggested that Newton recognised that the idea of absolute motion was operationally meaningless. Newton's dynamics became what it has been ever since—the science of causes of changes of motion. "To this science, even as it is presented in the *Principia*, the concept of absolute motion is utterly without consequence." Why then did Newton defend the idea? "Possibly germane to the question is the increasing stridency of his assertions, which grew, both in vehemence and in length, in exact proportion as the development of his dynamics rendered the concept operationally meaningless." [Westfall, p.445.] If Westfall is right, it has important philosophical consequences. It would locate the Clarke-Leibniz correspondence in a predominantly theological or metaphysical context. Newton, we know, hated metaphysical disputes, especially public ones: he was quite content to leave the matter to Clarke. Leibniz was from the outset attacking Newton's theology. If Newton himself recognised that dynamically, absolute space and motion were operationally meaningless, then most of what divided him from Leibniz was the role of God in the scheme of Nature. Interestingly, Westfall expresses this in the context of a discussion of the development of Newton's concept of force. The latter's motives, Westfall suggests, can be traced to *De Gravitatione*, where absolute space expresses his revulsion from the absolute necessity of a world in which no guidelines and reference points were present. "'The eternal silence of these infinite spaces fills me with fear.' Pascal's *cri de coeur* found its echo in Newton's refusal to set sail on the shoreless sea of relativity. By vehemence alone, when all else failed, he would refuse the manifest conclusion to which his own dynamics led him. His assertion of absolute motion has all the appearance of an act of defiance hurled in the face of the very current of thought on which his dynamics itself was borne inexorably towards its ultimate form." [Westfall, p.446.]¹¹

Newton has shown that certain effects demand an explanation. He believes he has also shown that the alternative viewpoint cannot account for all the facts. Certain motions imply that a force is acting on a body; this differentiates real from apparent motion, since there can be relative motions in the absence of such forces. Even Leibniz recognised that certain motions are distinguishable by means of the presence of the "cause" being in one body rather than another. Indeed, the concept of force is basic and, in a certain sense absolute as an explanatory category, for both Newton and Leibniz. For Newton, forces are the causes of certain motions which can only be explained fully by postulating absolute space.¹² For Leibniz, force—*vis viva*—is a characteristic that is fundamental to the nature of substance. The difference in the two notions is that Newton deals with their effects mathematically. As to the real nature of forces as causes, and as to an explanation of their ultimate character—on this, Newton is content to remain silent. Leibniz on the other hand will not content himself with merely "phenomenal" descriptions: force is an ultimate metaphysical category, and is capable of offering genuine explanations of the phenomenal world.

Alexander Koyré has suggested that the Newtonian discovery of the absolute character of rotation—in contradistinction to rectilinear translation—constitutes a decisive confirmation of his conception of absolute space. It makes it empirically accessible without depriving it of its function as a metaphysical postulate. Koyré's point is that absolute space is inevitable once space is "geometrised". If it is the inertial motion

---

¹¹ Against this, Pap has argued that '...any reference frame with respect to which the law of inertia is true...gives an operational meaning [to absolute space].' However, since the law of inertia is an idealisation, then absolute is that ideal reference frame with respect to which this ideal law is true. This is hardly what Newton demands of absolute space, and is much closer to the view that Kant would take of it, viz., that it is an Idea of Reason. (See Arthur Pap, p.18.)

¹² It is interesting that Kant distinguished absolute from true motion in a way that Leibniz did not. See Kant, *Metaphysical Foundations of Natural Science*, p.128.

that becomes, like rest, the natural status of a body, then the circular motion, which at any point of its path changes its direction, appears from the point of view of the law of inertia not as a uniform, but as a constantly accelerated motion. But acceleration has always been absolute, and remained so until Einstein. Yet, "as Einstein reclosed the universe, and denied the Euclidean structure of space, he has, by this fact, confirmed the correctness of the Newtonian conception." [Koyré, p.169.] This relates to the connection between space as absolute, and space as Euclidean, infinite, and homogeneous. But as I suggested above, Newton, in asserting the absolute character of space and denying the relational view, is committed to the idea that there are no well-defined positions in space; that is, he is denying that space is an infinite undifferentiated whole. For Newton, absolute space has a "centre", and all points in it are well-defined in principle and are genuinely different.

This raises profound problems of course. If all points of space are unique such that the "world-line" of any particle is well-defined, and if points of space can be re-identified, then the problem of defining relative motion becomes acute. My desk may be at rest relative to the walls of my study, but since it shares the motion of the earth around the sun it must successively occupy different points of space. Clearly my desk must always occupy some point of absolute space—there is nowhere that a body could be except in some point of absolute space, and at some point of absolute time. Thus all bodies that are moving only relatively to some frame of reference which is itself in uniform motion, are moving successively through points in absolute space. We might be at rest—i.e., occupying the same point of absolute space through a period of time—relative to the rest of the universe. But accelerated motions tell us that forces are acting; and where forces are acting there is change of state. All motion becomes observable only by reference to other bodies—the "lattice" of space is ideal, not real—but forces cause changes, and changes occur even when there are no reference bodies against which to compare them. By making rest and uniform motion "states", Newton has made it essential to understand change, and thus force, as fundamental categories of rational dynamics. (The use of calculus also strengthens the idea that motion can be treated as if it were a state.)

What of absolute time? Newton has claimed to have shown that the relational theory of space is inadequate to a full explanation of the facts, and that his own theory follows from both physical and metaphysical considerations. There is no doubt that the idea of absolute time does not have a comparable base in experience. Does absolute time require independent justification? Clearly all motions, real and apparent, with which Newton is concerned, take place in time. There is only one time—relative times being simply inadequate measures of it. And in the same way that all bodies have a place in the one all-encompassing space, so all things, processes, changes etc., occupy successive moments of absolute time. Absolute motion is the translation of a body from one absolute place to another; relative motion is a translation from one relative place to another. Both translations take place in absolute time. Relative spaces are parts of absolute space and, similarly, relative times are parts of absolute time. Ernst Mach complained that absolute time had neither a practical nor a scientific value, though Newton gave it both when talking about the reference frame used when assessing the accuracy of clocks against celestial motions. Nevertheless, there does seem to be, in the application of measures to periodic processes, an assumption to the effect that our measures are at best approximate; such an idea seems to involve the possibility, at least in principle, that these measures could be made exact. Nonetheless, it would be hard to make intelligible the idea that we had in reality found a measure of absolute time as such; but we imply, when we measure time against one process rather than another, that the first is in some undefined sense closer to the hypothetical or "ideal" timekeeper. But in the case of Newtonian time, we do not know if our measurements are in principle relative, or if the possibility exists of finding an exact measure of this time. It is difficult to make such a possibility coherent; any measure of time is logically independent of that which it measures. The idea of finding "time itself" seems meaningless.

Thus space and time are not perfect measures, but are that against which our measures are judged. No actual reference frame could be identified with Newton's "absolutes": they function as Ideas of Reason (in Kant's sense) giving metaphysical unity to the Newtonian universe of matter in motion. It is the separation of space and time from their basis in experience that leads to profound difficulties for Newton's theory. His problems are ultimately concerned with motion: true and apparent motion can be distinguished by means of the various forces that Newton discusses. But motion always involves both space and time: there is no metaphysical reason why we should feel the necessity to move from the existence of different types of motion, to the existence of different types of space. The fact that Newton offers no separate arguments to show the existence of absolute time should be seen as due the fact that time and space belong together whenever any kind of motion is under discussion. If Newton's arguments really do demonstrate that from absolute motion we can infer absolute space, then they should in turn demonstrate that we can infer absolute time. If motion involves both space and time, then any premises that contain the idea of motion must lead to conclusions about space and time, and not merely about space. Newton does not seem to have contemplated the possibility that he might have accepted the existence of absolute motion without absolute space. It seems to me that the reason for this is to be found in his theological presuppositions. Even Leibniz came to accept that there was a real difference between motions, dependent upon the location of the "cause".

So far as space and time are concerned, Newton has left his "positivism" far behind. Absolute space is not so much operationally meaningless, as operationally redundant. These concepts are not given in experience: do they then have any physical significance? The laws of natural science, considered as inductive generalisations, must be placed within some theoretical structure which itself transcends sense-experience. Science is not simply a collection of facts. Organisational concepts, theoretical constructs, have an essential function even in experimental natural philosophy. Newton certainly recognises this, even if he very often excludes any discussion of the nature of the fundamental metaphysical presuppositions which make any world-picture possible. In this sense it is simplistic to regard Newton even as an empiricist; if empiricism is intended as an explanation of the phenomenal world, then Newton is no empiricist. For him, as for Kant, empiricism is a programme for science, a means of describing the entities with which the natural scientist deals. The "ultimate" explanation is outside the scope of this programme, and what we say about it must remain hypothetical:

> ...the main business of natural philosophy is to argue from phenomena without feigning hypotheses, and to deduce causes from effects, till we come to the very first cause, which certainly is not mechanical...And though every step made in this philosophy brings us not immediately to the knowledge of the first cause, yet it brings us nearer to it, and on that account is to be highly valued. [In H.G.Alexander, p.174.]

There are really two distinct Newtonian methods. First, there is the view that science consists of mathematical laws stating the behaviour of nature, laws deducible from and verified by phenomena. Second, there is the ideal and unattainable explanation of these phenomena, i.e., metaphysics. Absolute space and time are not part of, and cannot belong inside, the first framework; but they are necessary conditions for the mathematical idealisations which make this method possible. So far as perceptual space and time are concerned, Newton denies none of the phenomenological facts. They are merely relegated to the status of "vulgar prejudices"—or rather, they give rise to such prejudices as taking the real time for our imperfect measures of them. They are derivative facets of real, absolute space and time: they are real, but not ultimate.

The tensions between the two sides of Newton's method are more obvious to us than they could ever have been at the time—although the attacks from both Leibniz and Berkeley demonstrate forcibly that not everyone was convinced—when it was not automatically regarded as irrelevant to include the nature of God in a supposedly scientific scheme of nature. Theology and science are closely interwoven in Newton's thinking on space and time, but the nature of these concepts as "effects" of God—as a *sensorium Dei*—are not logical deductions from phenomena. Would Newton have thought it possible to prove the existence of God's attributes by means of an experiment with buckets and ropes? More likely, Newton was concerned with such arguments to point out the difficulties of a relational view of space and time. What then follows is an analogical leap to absolute space and time, which are accepted as real by Newton prior to these arguments, but accepted largely for theoretical or, as Westfall argues, psychological, reasons: "To avoid the Scylla of relativity, Newton embraced the Charybdis of absolute space." [Westfall, p.445.]

1.2 Leibniz: The Ideal and the Real

Whereas for Newton there must be, in the nature of the case, certain areas forever outside the scope of human knowledge and, in particular, beyond scientific explanation; for Leibniz, the ultimate nature of the universe is transparent to reason. Metaphysical principles can provide genuine explanations of phenomena and their foundations in ultimate reality. Indeed, for Leibniz the phenomenal world is in a certain sense merely the confused perception that finite minds have of the fundamental world of monads and their states. The debate that Leibniz has with Clarke is essentially about the nature of space and time, and the interpretation to be given to metaphysical principles which, he believes, are really shared by his opponents, but have not been properly understood or rigorously applied.

The correspondence with Clarke offers a sustained and systematic exposition of some of Leibniz's most important views on space and time. Leibniz is occupied from the outset with what he calls the "decay of natural religion" in England; a decay which he attributes to the predominance of philosophical views based on Newton. For Leibniz, the writings of Locke and Newton in particular seem designed to diminish God's role in the world.

In the early correspondence with Clarke, Leibniz is more concerned to refute the Newtonian conception of space and time than he is to present an alternative. His main objection to the Newtonian model is that both absolute space and time violate the "supreme" Principle of Sufficient Reason. Among the many complex ways in which this principle can be described, perhaps the one most important for our purposes is the logical thesis that every true proposition is analytic. [See Broad, (1975), p.11.] In Leibniz's opinion, this does not remove the distinction between the necessary and the contingent, since the analyticity involved is "infinite" in the case of contingent propositions. (In other words, only God knows all of the predicates which make a "contingent" proposition true.)

This logico-metaphysical principle has ramifications throughout Leibniz's philosophy. It takes us directly to his idea that every predicate of a true proposition is "contained in" the notion of the subject of which it is the predicate. That is, all states of a substance—past, present, and future—are "contained in" the subject's complete individual notion, which God sees as a totality prior to the actual unfolding of these states.[13] This connection between predicate and subject leads to the most general formulation of the principle of sufficient reason, which states that for every contingent fact there is a reason why it is thus rather than otherwise. In other words, all truths are grounded in the nature of things, and nothing whatever happens without there being a sufficient reason for it. Put like this, the principle looks metaphysical/causal in nature: it seems often to be seen as such by Leibniz. In the early letters to Clarke however, the principle is seen not so much as a logical thesis regarding the link between subject-concepts and their predicates—the "propositional link" as Buchdahl calls it—but as a principle of what we might call "possibility actualisation" binding even on God. If it is

---

[13] This principle has both logical and ethical connotations for Leibniz. For an extended discussion of Leibniz's principles in relation to the philosophy of science in particular, see Buchdahl, (1969), especially Chapter VII.

assumed that absolute space exists then, due to its homogeneity, no point in it differs from any other point. Therefore if God considered actualising, say, two possible substances, there could be nothing "internal" to the complete individual notions of either which could correspond, or express, their unique position in space. If God has a genuine choice of where to place the bodies, this position in absolute space would have to be a fact about each body, and would be ultimately derivable from their complete individual notions. Since absolute space thus makes no difference to anything which could possibly be a unique state of each body, their position in such a space is a difference which makes no difference. Thus God has no real "choice" in such a case— *agendo nihil agere*. But since there are bodies, either there must have been a genuine choice, or absolute space cannot exist. Leibniz opts for the latter, since the former implies a rejection of the uniformity and homogeneity of space. He will certainly not entertain the suggestion made by Clarke that the "mere will" of God is a sufficient reason.

Leibniz rejects the very idea of void space: this, he thinks, would merely be a special case of absolute space, and furthermore would also violate another metaphysical principle he accepts as true, viz. the "principle of perfection". An immediate consequence of this is the "principle of plenitude". Leibniz asserts that there could be no possible reason, no sufficient reason, for limiting the amount of matter in the universe, for the more matter there is, the more opportunity God has for exercising His wisdom and power. But Leibniz is ambiguous on this point: he frequently writes that if God wishes to create the best possible world, then this directly implies a "full" universe. However, he neglects his own principle that the maximum perfection is gained when God has realised the greatest variety of things compatible with order: a disordered universe would be imperfect. This principle—perfection= variety+order—is usually neglected by Leibniz in favour of the principle of plenitude *simpliciter*, i.e. perfection=the physical plenum. [Leibniz: in Latta, p.249.] Since for Leibniz the principle of perfection implies the principle of plenitude, whenever God can place matter somewhere, He will do so. If we assume empty space, it follows that God could have placed matter in it; therefore, wishing to maximise perfection, He has done so. Thus God has placed matter in any empty space, and in all empty space: therefore the universe is full. The argument is only valid for Leibniz when he operates with the "plenitude" version of the principle of perfection. Taking the first version, Leibniz might concede the possibility of empty space, for once God reaches a certain point with created matter, the further actualisation of possible substances would be, so to speak, held up, if this were likely to lead to a less ordered universe. One additional consideration here might be that the laws of motion must be more complex in a plenum than in a void—Newton had already pointed out the serious difficulties in trying to account for the phenomenon of gravitation in a universe full of matter arranged in Cartesian vortices.

Clarke objects that even on the assumption of relational space there is still no sufficient reason for God to place three equal particles in one order rather than another. Leibniz argues that Clarke must make some independent sense of the idea that there could be three exactly equal particles, without thereby begging the question against the principle of the identity of indiscernibles; and also that there may indeed be a difference dynamically, rather than in terms of the "order of situations".

The argument that Clarke uses to most effect against the doctrine of relational space attacks that position where it would seem to be strongest. Leibniz insists that where there are no observable consequences, then, in principle, the assertion of differences must be without significance. Clarke, taking his lead from Newton, uses the arguments from centrifugal force to prove that motion relative to other objects, and motion relative to absolute space, can be dynamically distinguished. Not all of Clarke's examples are damaging to Leibniz, but there can be little doubt that the latter is in a difficult position with respect to these arguments. The first of Clarke's arguments comes in the *Third Reply*:

> If space was nothing but the order of things coexisting; it would follow, that if God should remove in a straight line the whole material world entire, with any

swiftness whatsoever; yet it would still always continue in the same place; and that nothing would receive any shock upon the most sudden stopping of that motion. [In H.G.Alexander, p.32.]

In this example Clarke is treating the whole material universe as though it was itself a partial system capable of being acted upon by some outside force—in this case, God—which is not part of the system. Leibniz could appeal to his thesis of the equivalence of hypotheses: so far as the relative motion is concerned—that is, the change of situation—there is no way of distinguishing the case where a force acts on the earth, thereby causing objects to move, from a case where the objects are accelerated and the earth continues its uniform motion. Clarke is begging the question against Leibniz by specifying that the force acts on the earth; on a Leibnizian view, the ultimate cause of the change of relative situation may be unknown. However, Leibniz's actual reply to this point is not along these lines. It probably illustrates that he has not fully grasped the significance of the argument from forces. He says that

[T]o say that God can cause the whole universe to move forward in a right line, or in any other line, without making otherwise any alteration in it; is another chimerical supposition. For two states indiscernible from each other, are the same state. [In H.G.Alexander, p.38.]

What Clarke was endeavouring to show was that the two states are discernible; what he should have pointed out here is that though two states—uniform motion in a straight line, and rest with respect to space—are indiscernible while no changes take place; nonetheless, if we were to observe the change of relative situation we could then say that a force has been applied, and thus that either the earth or the bodies on it are in absolute motion.

Leibniz fares better with Clarke's example with time. Clarke says that unless we assume absolute time as independent of events, we would have to say that if God had created the world millions of years sooner than He did, "...yet it would not have been created at all the sooner". By appealing to the principle of perfection, Leibniz is able to point out that if there was a time before the creation of the material world, then God would have neglected an opportunity for maximising perfection; thus the world was created "...before any assignable time, that is, the world is eternal".

Clarke goes on the offensive over the same point in his *Fourth Reply*; he makes it clear that uniform motion and rest are both states, and changes of state lead to observable effects. The argument, he insists, "...is a mathematical one; showing, from real effect, that there may be a real motion where there is none relative; and relative motion, where there is none real". [In H.G.Alexander, p.48.] Once again, Leibniz can take a "positivist" stand: the concept of motion is meaningful only from within some reference system, and all such motion is phenomenologically or kinematically relative.

I now want to argue for the idea that Leibnizian spatial and temporal relativism can be properly understood only when the various levels of reality in Leibniz's metaphysics have been clearly distinguished. Most importantly, I believe that a modification is demanded of one influential view, viz., the interpretation that Leibnizian space and time are *phenomena bene fundata*. This view, (associated in particular with Rescher), places unwarranted emphasis on the epistemology of Leibnizian space and time, unnecessarily disregarding significant ontological considerations. What I shall argue is that this characterisation of Leibniz's view is, if not strictly false, then seriously misleading. It seems to me that one reason for the acceptance of this position is that Leibniz frequently addresses himself to his adversaries on different levels; and though each level of discourse may be internally appropriate, it can mislead, if it is then extrapolated to the other levels in Leibniz's highly complex system. [See Buchdahl, 1969, p.465]

There is clearly a place for space and time locutions on all three levels of Leibnizian metaphysics. There is firstly the ontological level—the "ultimate" reality of the monads and their states. Secondly, there is then the epistemological level—the "derived" reality of phenomenal aggregates. And thirdly, there is the "merely" intellectual reality of *entia rationis*, which includes determinate space and time qua idealisations. The traffic between these levels is inevitably complicated, and the coherence (or lack of it) of this hierarchical system is to an important extent given its crucial exemplification in the explanation of space and time.

An additional complication for the interpreter is that the correspondence with Clarke can lead to the assumption that what Leibniz says therein may be taken as a definitive view on space and time in particular. In fact, in that correspondence, Leibniz nowhere refers to space and time as *phenomena bene fundata*. Nevertheless, commentators sometimes seem to take what Leibniz says there as definitive, particularly in that important section where Leibniz discusses the question of how the idea of space is acquired. If this view concerning the acquisition of the *idea* of space and time is then linked to the view that in one respect at least space and time are *phenomena bene fundata*, some of Leibniz's ideas on this important subject are made to seem irremediably obscure or, if not obscure, then inconsistent. In what follows I hope to disentangle some of the confusion that results from a conflation of the different levels within Leibniz's system.

In the correspondence with Clarke, Leibniz's major preoccupation is the refutation of Newtonian space and time. This is the essential contextual framework for the debate, and it is important to keep this in mind in any consideration of what Leibniz says there. It is misleading to take what he says there as any statement of a general metaphysics of space and time. The views there need to be fleshed out in the light of Leibniz's wider concerns.

In describing the manner by which we arrive at the concept of space Leibniz asks us to consider, firstly, "the existence of many things at once." We observe in them a certain "order of co-existence". This order is their situation or distance. What Leibniz will try to show is that the qualitative notion of an "order of co-existence" *precedes* the idea of space as quantum. This is the key to Leibniz's whole theory of space and time. Failure to understand this has led to the view—originating in the correspondence with Clarke, and receiving fresh impetus in Kant—that Leibniz's position is question-begging. Kant, for example, says that:

> It is only through the idea of time that it is possible for the things which come before the senses to be represented as being simultaneous or coming after one another. Nor does succession generate the concept of time, but it makes appeal to it. And so the notion of time regarded as though acquired through experience is very badly defined, when it is defined by means of the series of actual things which exist one *after* the other. For I do not understand the meaning of the little word *after*, except by means of the already previous concept of time. For those things come *after* one another which exist at *different times*, just as those things are *simultaneously* which exist *at the same time*. [Kant, P.C., p.63][14]

Leibniz claims that a purely relational interpretation of space and time admits of being quantified. He writes: "Order also has its quantity; there is in it, that which goes before and that which follows; there is distance or interval. Relative things have their quantity, as well as absolute ones. For instance ratios or proportions in mathematics, have their quantity, and are measured in logarithms, and yet they are relations." [In H.G.Alexander, p.75]. Leibniz defines distance as a serial ordering of before and after; distance and "betweenness" are characterised by a non-metrical ordering relation. One of these co-existing bodies may then change its situation relative to the others, while the relative situations among the other bodies remains the same. Given bodies A, B, C; then A changes its relationship relative to B, C; while the latter remain in the same situation relative to one another. Suppose then that another body, D, takes up the situation, relative to B, C, that A previously had. We now say that D is in the same place once occupied by A. Furthermore, if the cause of the change of relative situation is "in" A, then we can say that A has "really" moved. Of course we may not know if the cause is in A; we must then leave open the question of whether A was in motion, or B and C together. In the latter case, we could say that B and C, while preserving their relative situations

---

[14] For Kant's detailed criticisms of Leibnizian metaphysics, including the theory of space and time, see *Critique of Pure Reason*, especially A261/B316ff.

one to another, had left the vicinity of D, which then occupied the same relative situation that A had done. [In H.G.Alexander, p.69-70].

Using this simple analysis, Leibniz continues by defining "place" in terms of "same place" by means of an equivalence relation. And that which "comprehends" all those places taken together, we call space. The term "space" is thus merely a convenient term denoting the universe of discourse of all these actual and possible relations of situation. The idea of "same place" does not presuppose an independently defined notion of place. Defining place—and hence space—in this fashion, renders the concept of absolute space qua container existing prior to and independently of objects, otiose. Reference to the place of a body is simply a useful way of referring to the relations that a body has to those bodies around it.

Leibniz insists that we should not hypostatise these places: this would commit us to the substantival space of Newton which such an analysis is designed to avoid; it would mean confusing the repetition of a relation with the existence of a thing. [See Cassirer, 1902, p.254]. Place, for Leibniz, is that which is the same in different moments to different existents when their relations of co-existence—assumed to continue fixed from one of these moments to another—agree entirely together. Fixed existents are those in which there has been no cause of any change of the order of their co-existence with others, or in other words in which there has been no motion.

We must further distinguish "place" from the relation of situation which is "in" the body that fills the place. The place of A and D is the same, but the relation of situation that first A, and then D, has to B and C, is not the same. Two "accidents" cannot be in one subject, and the relation here is a genuine "affection" or accident of the two bodies. Clarke had claimed against Leibniz that "...it would be absolutely indifferent, and there could be no other reason than mere will, why three equal particles should be placed or ranged in the order *a, b, c,* rather than the contrary order." [In H.G.Alexander, p.30.] Leibniz's reply to this is an appeal to the principle of sufficient reason, insisting that if these alternatives were really indifferent, God would not create them at all. Now this, it seems to me, is an example of Leibniz responding in the less than rigorous terms he frequently allows himself in his letters to Clarke. He could have responded—with more plausibility—that though the places of the three particles may be the same in each case, their relations of situation are not. Since the latter are "affections" of the bodies at particular moments of their history, there is a genuine difference in the two cases, viz., as part of the respective monadic perceptual histories. For Clarke is assuming not only that the bodies are identical, but also that the places of them are undifferentiated. In the discussion of place there is no suggestion, (*pace* Rescher), that Leibniz held all relational predicates to be ultimately reducible to a series of conjunctions of non-relational (monadic) predicates.[15] On this level of reality, Leibniz affirms the irreducibly relational character of certain predicates. We could reduce the notion of space to non-relational predicates only if the well-foundedness of space was based upon the places of the bodies. But the "place" of a body is itself an idealisation, a figure of speech, with respect to the relative situations of co-existing bodies. In so far as the concept of space is based upon the character of bodies it is based, ultimately, on the relations of situation that each body has to the rest of the universe; and is thereby based on relational predicates. Leibniz says that the mind, "...not contented with an agreement, looks for an identity, for something that should be truly the same." [In H.G.Alexander, p.70.] That is, the relation that A has to B and C, "agrees with" the relation D has to B, C. Extrapolating this actual relation into a possible relation for all bodies whatever yields an ideal system of actual and possible relations that is conceived as extrinsic to bodies; this is what we call place, and space. Yet *this can only be an ideal thing.* [Ibid.,p.89.]

Leibniz offers two well-known examples as illustrations. The relations we call spatial are ideal in the same sense in which the lines of a family tree are ideal: and we make space into a thing when we confuse the map

---

[15] See Rescher, (1967), p.74 ff; and Hidé Ishiguro, in H.Frankfurt, (1976), p. 155-90, and 191-213.

with the territory. The lines linking fathers and sons on a family tree are obviously not real in the sense in which the fathers and sons themselves are real—there is no "entity" in the world called "paternity". His second example anticipates a conventionalist theory of measurement. In this case Leibniz does not assert the unreality of the relation, but merely that measurement presupposes a point of view; that is, that the relation is different depending on the perspective taken on the entities related. Thus: "In the first way of considering [the two lines] L the greater, in the second M the lesser, is the subject of that accident which philosophers call relation."[16] In the third way of considering the two lines, viz., as a proportion, we do not state which of them is the "subject" and which the "object"; this relation is therefore "out of the subjects". Since it is neither a subject nor an attribute it must, in Leibniz's terms, be merely an "ideal" thing. Relations like this can be expressed in subject-predicate propositions, asserting an attribute as belonging to a subject. If the relation is considered as a proportion it is deemed merely "ideal". The tendency to hypostatise ideal relations into substantial things is strengthened further by the use of certain analogies: we perceive, for instance, the trace left by a moving object, and then imagine such an immovable background entity such as "pure space" for motion in general. "But this is a mere ideal thing, and imports only, that if there was any unmoved thing there, the trace might be marked out upon it." [Ibid.,p.72.]

Leibniz's analogy of the family tree illustrates his point about the idea of space tolerably well. But what of time? I suggest that we might usefully employ the idea of intervals in music to illustrate the Leibnizian notion of time as an order of successions. The notion of a musical interval will serve rather well to illustrate ideal temporal relations which are yet "well-founded". A musical interval has no "existence" outside the notes whose relationship the interval expresses: the notes are real enough, but the

---

[16] Leibniz, in H.G.Alexander, p.71. This is one of the most unambiguous examples of Leibniz's acceptance of relational predicates.

interval is a measure of the relationship the notes have, qualitatively, to one another. Yet although it is the quality that each note possesses—its pitch—which determines the interval, the interval as such is quantitative in musical terms. The quality of each note is expressed on an ordered scale of harmonic relationships; it is a measure of the "positions" the respective notes have on a musical scale. By analogy, distance can be regarded as a serial ordering capable of expression in terms of numerical proportions. Taking Leibniz's substances and their states as the "notes", the successive unfolding of states—considered "across" substances, so to speak—yields temporal intervals, in this case "public" time. The one-directional unfolding of the respective complete notions of substances makes the "arrow" of time itself one-directional.

Now it is at just this point that it becomes easy to confuse the different levels of reality in Leibniz's system that I distinguished above. Space and time may be regarded as *phenomena bene fundata*, but we must be careful to draw certain distinctions without which Leibniz's assertions in the correspondence with Clarke are bound to remain obscure. The view that space and time are phenomena must be seen in the context of Leibniz's whole metaphysical system. For Leibniz to say that something is "merely phenomenal" and yet *bene fundatum* means, that though it is not ultimately real—like the monads—it is nevertheless capable of being coherently experienced. Space and time are *bene fundata* by virtue of the fact that they constitute a coherent system of ordering relations of phenomenal aggregates to which we may attribute the qualities of extension and change. If space and time were merely illusions, this would be impossible. Phenomenal space and time become known by means of repetitions—repetitions of "contiguity" and "succession". The pre-established harmony ensures that to every state of a monad at some instant of its private, "internal" time, there corresponds a state of every other monad contemporaneous with it. Such a system of inter-monadic simultaneous states yields "public" time. Thus for Leibniz, the idea of nature "at an instant", i.e., absolute simultaneity, is well-defined at the phenomenal level, and guaranteed at the monadic level by the pre-established harmony.

Leibniz offers no analysis of time comparable to that given for space. This is made clear, implicitly, by Chana Cox. For in spite of the plausible reconstruction Cox offers of Leibniz's spatial relativism, some of those key ideas cannot easily be extended to embrace temporal relativism. It seems to me that a metaphysical system must be able to deal consistently and analogously with both space and time—and it also seems to me that Leibniz's metaphysics begins to fracture at just this point. [Cox, p.87-111.]

Phenomenal time is logically prior to space: to develop the idea of space we assume the co-existence of a number of bodies *in* space. The concept of place is reached by means of the idea of "same place"; this is then generalised to embrace all possible places *as if* it were a system of possible as well as actual locations. Yet the essence of time is succession: we cannot employ an equivalence relation and reach the idea of time by this means. We can generalise the idea of particular places into the idea of space, but time is not equivalently a generalisation from particular times. The "before and after" relation of moments of time is asymmetrical, and cannot therefore form the basis of an equivalence relation, which requires transitivity, reflexivity, and symmetry. The directedness of time cannot be reduced to a series of relations of "same time" without loss of this essential quality. For although "x is simultaneous with y" is an equivalence relation, "before and after" is not. In the Newtonian system, two successive moments of time remain distinguishable even if all physical events occupying these moments are identical in all respects. For Leibniz, however, the relation of succession is itself characterised by physical differences—viz., changes of state, and each substance is subject to a perpetual, continuous such change of state, i.e., appetition.

The particular phenomenal status that space and time have is a function of the states of the monads. If the mutual perceptions of the monads did not agree, the space and time given to us as phenomena would be different. For example, if some monads did not mirror all the others but only some of them—perhaps those "closest" to them—the system of co-ordinated perceptions would be split into spatially unrelated sub-systems, each one of which might be, as such, spatially co-ordinated. A "chaos" with no spatial structure, (no phenomenal structure, that is), would consist of a "system" of self-contained monads with no perceptions outside of their own internal states. Similarly, if the appetitions of the monadic programmes did not have the required coherence—guaranteed by the pre-established harmony—then public, inter-monadic time would not exist. And if the changes of state of the monads was continuous, but not in harmony, we might have two or more "times", each of which would be *bene fundatum* but phenomenally unrelated.[17]

Two further comments are apposite here. Leibniz says that "place" is that which is the same in different moments to different existents, when their relations of co-existence to certain other existents "agree entirely together". Now Russell suggests that when Leibniz says that these other existents "...are supposed to continue fixed from one of these moments to the other...", then he (Leibniz) is making a supposition which on the relational theory is "...wholly and absolutely devoid of meaning." [Russell, p.121.] Against Russell's objection, it need only be emphasised that a system of reference for motion, or "change of relative situation", is essential for both absolute and relational space. Leibniz insists that motion, in order to be observable in principle and thus to involve a "real" change, involves the idea of one body changing its situation relative to some other body or bodies which retain their relative situations with respect to each other. The idea is perhaps clearer if we emphasise that the other bodies, against which the motion is deemed to occur, are *supposed* to continue fixed: that is, we consider these bodies *as if* they were fixed, and thereby provide a reference frame for the motion.

---

[17] Some of these interesting—and probably un-Leibnizian—possibilities, are suggested by Rescher, (1967), p.90.

CHAPTER ONE – PROLOGUE: NEWTON AND LEIBNIZ

To understand Leibniz's mature theory we must distinguish three levels on which it operates.[18] First, there is the ultimately real level of the monads; second, there is the level of phenomena, which depends on the monads and their states; and third, there is the level of the ideal, the "useful fiction", or *ens rationis*. The analysis of space given in the *Fifth Paper* is of the possibility of determinate space given the assumption of extended phenomenal aggregates. Space, as *phenomenon bene fundatum*, is the appearance of coherently ordered aggregates which depend upon the modes of genuinely real substances. Leibniz's analysis in terms of co-existing bodies appears circular only so long as it is assumed that he wishes to explain how we come by the idea of *spatiality*—which is really for him already and always *phenomenon bene fundatum*—rather than space—determinate space—which is ideal.[19] Abstract space and time are more like the entities of pure mathematics; that is, they are abstractions, or *ens rationis*. As Leibniz points out, he does not say that space "...is an *order* or *situation*, which makes things capable of being situated: this would be nonsense. Space is [not] an order or situation, but an *order of situations* when they are conceived as being possible." [In H.G.Alexander, p.89, my emphasis.]

Leibniz's account of the acquisition of the concept of space would be viciously circular, as Kant seems to have thought, only if his analysis did not take cognisance of the fact that our perceptions are of an already spatially and temporally extended world of monadic aggregates. The mind forms an abstract idea of space and time which embraces the already given relations of situation: this, extrapolated to abstract possible relations, yields and *idea* of space which obviates the necessity of postulating a real entity, like absolute space, which exists independently of the objects in it. It is the assertion of metric relations among bodies that Leibniz regards as a "fiction": "space" is a fiction—spatiality is a

---

[18] For an elaboration of this idea, not concerning space and time in particular, but for Leibniz's metaphysics as a whole, see J.E.McGuire, p.290-326.
[19] For this useful locution, see Buchdahl, (1969), p.573, note 2; and Chapter VIII, Section 7, *passim*.

phenomenon bene fundatum. As Leibniz puts it: "There is no nearness or distance, whether spatial or absolute, among monads, and to *say* that they are collected together in one point, or dispersed throughout space, is to make use of certain fictions of our mind, by which we try to represent to ourselves in imagination what cannot be imagined but only understood." [Leibniz, edited by Gerhardt, ii, p.450.]

Leibniz's system, then, requires three levels. Space and time, as analysed in the correspondence with Clarke are two stages removed from ultimate reality: *phenomena bene fundata* are one stage removed from the ultimately real. We might still say that space and time are ultimately founded in the monads and their states through the mediation of phenomenal aggregates and their relations. [See Loemker, p.264.] Thus space and time could be characterised as *"fictiones bene fundata"*, though as far as I know this term does not occur in such a context in Leibniz's writings. Given this tripartite metaphysical scheme, it is not really surprising that crucial problems of interpretation arise out of Leibniz's analysis of the concept of continuity, a concept that finds application on more than one level. As Leibniz writes: "There are two labyrinths in which the human mind is caught. One concerns the composition of the continuum; the other concerns the nature of freedom. And both arise from the same source, namely, the infinite." [In Loemker, p.270.] We see an interesting parallel here with Kant. The problems of infinite divisibility, the continuum, Ideas of Reason, and so on, are also regarded by Kant as a fecund source of paradoxes and confusions, not resolved until the solutions of Transcendental Idealism are applied to the Antinomies of Pure Reason.

Leibniz rejects the notion of the actual infinite, holding that a definition must involve a proof of the possibility of the thing defined. This denial of infinite number means that Leibniz is required to say something positive about the supposed existence of infinitely small numbers, as well as the assumed infinity of created substances. He avoids the first problem by denying the ultimate reality of infinitesimals: these are relegated to the status of useful fictions. The answer to the second problem turns on the application of certain theoretical concepts to experience. Since the universe of created substances exists prior to all composition, it is not formed by the successive addition of parts. The "true" infinity exists only in the absolute. Real unities are logically prior to their parts and cannot be resolved into them. This is what is implied by Leibniz's distinction of "resolution into notions", and "division into parts". He affirms the scholastic idea that "Being" and "Unity" are convertible terms. Thus only monads are true unities: they are unique in being both real and simple. Phenomenal aggregates lack this unity, which means that their unity is merely apparent, not metaphysical.

In general terms, Leibniz's solution of the continuum problem appeals to a distinction between metaphysical and abstract interpretations of the part-whole relationship. For example, a point is not a constitutive part of a line: the line is prior to its points, since the whole is prior to the parts. The "part" is possible, and ideal; the continuum is not divided into points, because *points are not parts but limits*. [In Gerhardt, iv., p.419.] Since a plurality of monads constitutes aggregates having only phenomenal reality, such a plurality yields only a "phenomenal" continuum. And it is the confounding of the real with the ideal which generates the kinds of problems crystallised in Zeno's paradoxes: "Those who compose the line with points have looked for first elements in ideal things, or for connections of a completely inappropriate kind." [In H.G.Alexander, p.63.]

This distinction is of great importance in Leibniz's philosophy. It explains why instants are not parts of time, nor mathematical points parts of the spatial continuum. [In Latta, p.300.] The solution to the problem, as Leibniz sees it, rests on a distinction between the actual and the potential, with its analogous applications in the realms of the real and the phenomenal. The terms of an infinite series constitute only a potential infinite, while the infinitely large and small belong to the actual infinite. For Leibniz, as for Kant, mathematical points are the boundaries of lines, and lines are the boundaries of surfaces. Points are indivisible just because there is nothing in them to divide. "Mathematical points are exact but they are only modalities...[they] are only extremities of the extended and modifications, of which it is certain that the *continuous* [*continuum*] cannot be composed." [In Latta, p.301.][20]

To summarise: only the monads are both real and simple; by aggregation these monads constitute the phenomenal continuum. Yet since phenomenal aggregates are not true wholes, they are not truly continuous. Space and time, as *entia rationis*, are continuous, but ideal. Points are not parts of space, and instants are not parts of time—the latter are modalities, ideal parts only, i.e., they are simple, but not real. Thus geometrical continuity belongs only to the abstractions of mathematics, and to determinate space and time.

Profound problems remain for Leibniz's theory. He has remarked on the continuity of phenomenal aggregates and on the geometrical continuum; yet he has neglected the putative continuous change of state of the monads through appetition. When Leibniz says that instants are not parts of time it must be remembered that time has been characterised either as *ens rationis*, or as "bare temporality", or as phenomenal change. Certainly, instants cannot be phenomenal; yet "phenomenal" is itself ambiguous. The time of physics is "phenomenal", yet so is the experienced time of minds. But whereas the time of physics might be regarded as mathematically continuous. the time of human experience is not, even though both such times are continuous in the unanalysed sense of not having "gaps". We might say that instants are not parts of phenomenal time qua the time of physics, since the application of the idealised mathematical continuum to the continuity of phenomenal reality is inappropriate.

---

[20] The problem of how metaphysical "points"—unextended foci of action—can constitute phenomenal aggregates that are spatially extended, may be answered intelligibly, if analogically, by appeal to the machinery of calculus. However, since the mathematical continuum, with its postulate of denseness, is an idealisation of the "intuitive" continuum, it would be going beyond anything that Leibniz suggests, to solve this problem by applying the properties of the mathematical continuum to the metaphysical continuity of monads.

Returning now to Leibniz's conception of time as the order of successions, it seems to me that there is a fundamental inconsistency in his treatment of temporality taken as phenomenon—that is, having a coherent structure—which is at the same time *bene fundatum*. According to Leibniz, we cannot speak of time in respect of the monads, yet we can speak of succession and duration. "Every thing has its own extension, its own duration, but it has not its own time, and does not keep its own space. Thus points are neither large nor small, and no leap is needed to pass them. Yet the continuous, though it has everywhere such indivisible points, is not composed of them." [In Gerhardt, i, p.416.]

While duration and extension are attributes of phenomenal aggregates, well-founded in the monads, time and space are external to both real monads and "mere", though well-founded, phenomena, and are the "measures" of them. I noted above that the pre-established harmony effectively guarantees for Leibniz the intelligibility of the idea of nature "at an instant" by co-ordinating the simultaneous states of all monads. But what kinds of states are these? They cannot be discrete because the complete individual notion of a monad is a metaphysical reality, an "eternal unity', and therefore cannot be composed of parts which are themselves real. The parts—that is, the states regarded as discrete—must then be ideal. Yet the pre-established harmony operates on the level of substances, ultimate realities, (although it has phenomenal implications), and is therefore assumed to be linking ultimate realities and not ideal notions. If we accept that appetition is really continuous, and that the totality of states exists, so to speak, "all of a piece" and prior to division; then there are an infinite number of "states" constituting each monad, and there are really no discrete states for the pre-established harmony to harmonise. Rather than regarding the complete notion of a monad as a series, perhaps like the number series, perhaps we might more appropriately regard particular times as "differentials", not fully determined and intelligible without reference to integrals—in this case the complete history of a monad. The duration of a substance would then not be understood as the permanency of a thing with temporally differentiable attributes, but as an "ordering law", the realisation of which expresses a persevering identity.

Leibniz's distinction of the real and the ideal separates perceptual time—*phenomenon bene fundatum*—from the time of physics and mathematics. The former is continuous, but not mathematically dense. Mathematical and physical time are idealisations and are not instantiated in experience. By denying that instants constitute time Leibniz has effectively removed the problems associated with infinite divisibility and succession. If time were composed of instants, no point of time would have a unique successor; this would then raise all of the paradoxes associated with Zeno that Leibniz is concerned to resolve. Problems arise whenever we picture time spatially—as a line, for instance—or in terms of a number series. As soon as we represent an instant of time by a number we transform that most characteristic temporal feature, viz., flux, as a series of discrete elements. When we picture the successive states of a monad as a series, say, S1 and S2 and S3...and Sn, we represent what is supposed to be continuous by a series which is discontinuous, and try to capture the essence of continuity by means of an infinite repetition of discontinuous elements. Leibniz works his way around this problem by denying that the monads have their own time and, more importantly, by attributing to the monads the notion of activity.[21]

It seems then, that successive changes of state do not, as it were, "add up" to the life history of a substance. Rather, the states of a monad, qua centre of force of continuous change, are aspects of this activity at any moment of phenomenal time. The spatial ordering of bodies as phenomenal aggregates is made possible by the dynamic relations of forces on the monadic level. The representation of the complete process by means of an individual element in it is possible only because we think of this individual as a special case of the comprehensive system of which

---

[21] On the level of monads and their states, this activity—this force—is "primitive" (*vis primitiva*); on the level of phenomenal aggregates it is manifested as a derived, mechanical force which is a phenomenal property of bodies. To assert that monads are foci of activity refers to the former concept. See Westfall, (1971), Chapter VI, *passim*.

it is part. Yet because, for Leibniz, the perception of order of succession is "confused", the particular spatial and temporal ordering that we do experience is not a necessary expression of monadic structure. Continuity on the perceptual level does not entail continuity on the monadic level, since the "confusion" that is supposedly characteristic of our perceptions does not directly entail that our perception are not orderly. Although we cannot, strictly speaking, refer to "causes" in Leibniz's system, we might say that the succession of causes and effects on the monadic level produces the phenomenon of change. Contemporaneous states—absolute simultaneity—are states that cannot causally interact. The topology of Leibnizian time is based on a metaphysical conception of causality, even though the latter term must be understood in a Pickwickian sense.

Leibniz suggest that "distance" between events is explicable by reference to the number of successive states interposed: "If the time is greater, there will be more successive and like states interposed; and if it less there will be fewer." Now what can Leibniz mean by this assertion that different intervals of time are determined by the numbers of successive states interposed between the relata? It suggests that there is a one-one correspondence between "states" and "moments" of time: every state would then have one and only one instant of time associated with it. But a state of a monad is an aspect of its activity: appetition and perception require intervals of time in which to occur. If they did not, each "state" would be infinitesimally small and would be, in Leibniz's terms, a fiction. Change of state takes place continuously—a continuity characterised, though not exhausted, by the continuity of the real numbers. Thus between any two states (numbers) there are an infinite number of other states (numbers); but the succession, the transition, from one state to the next, is a real process for the monad, whereas any arbitrarily chosen number of the mathematical continuum has no unique successor. Each state must pass into the next: time is the order of successions. If, *per impossibile* for Leibniz, the dynamic sequence of states was discontinuous, phenomenal space and time would also be

47

discontinuous. Yet the idea of an experience of a discontinuous space and time is incoherent. Matter may behave discontinuously—for example, as described within quantum theory—but a discontinuous space-time system offers a more radical challenge to intelligibility. Leibniz may insist on the differentiability of "states": the sequence of states would be transitive and asymmetric, and thus isomorphic with some numerical system. From such an isomorphism one could construct a temporal metric—a fiction, but a well-founded one. Leibniz's view that space and time are *entia rationis* but not arbitrary, suggests that the sequence of states, generating spatiality and temporality as *phenomena bene fundata* makes the metric intrinsic: we simply have to count the "moments" of which phenomenal time is composed, and assign numerical magnitudes to them. If space and time are continuous, this "counting" is impossible, and the metric must be introduced.

I do not pretend that distinguishing the three levels of reality in Leibniz's metaphysics resolves all interpretative problems in this area. However, one additional positive result should be noted here, in that the analysis given above throws into relief Kant's transformation of Leibniz, for part of what Kant does is to shift the balance of the real and the phenomenal as I have differentiated them. The level of ultimate reality is retained by Kant as the "intelligible cause" or "ground" of the world of appearances. This world of monads—or "noumena" as Kant will call them—is non-spatial and non-temporal. For Leibniz, the mind imposes a determinate space-time structure on an already given world of phenomenal extension and change. For Kant, on the other hand, the givenness of the phenomenal world presupposes a spatio-temporal ordering of the manifold of sensation by the mind. Kant unties the essential link that connects the ultimate reality of Leibniz's monads to the perceptual world, then re-ties the latter to Leibniz's *entia rationis*, space and time, which become presuppositional forms (in their guise as forms of intuition), that first make knowledge of the given world of appearances possible.

CHAPTER ONE – PROLOGUE: NEWTON AND LEIBNIZ

In Clarke's *Fifth Reply*, he once again returns to the attack, emphasising those points that he thinks Leibniz has not answered. The argument from centrifugal force is employed again, since Clarke is convinced that it effectively rebuts the relationist claims at that point where the latter theory is supposed to be strongest, viz., on the question of observable effects. Clarke writes:

> Neither is it sufficient barely to repeat [Leibniz's] assertion, that the motion of a finite material universe would be nothing, and (for want of other bodies to compare it with) would produce no discoverable change: unless he could disprove the instance which I gave of a very great change that would happen; viz. that the parts would be sensibly shocked by a sudden acceleration, or stopping of the motion of the whole. [In H.G.Alexander, p.104.]

What Clarke fails to understand is the far-reaching implications of Leibniz's positivist stand over space, time, and motion. Only the monads and their states are ultimately real; everything else—space, time, and motion—consists in phenomenal relations and have only ideal existence. Nevertheless, Leibniz's replies to the problems of differentiating absolute from relative motions appears to concede that there is a difference, and that this difference has to do with the "cause" of the motion, by which he must mean the existence of forces acting on some bodies rather than others. Actually, "acting on" others is a thoroughly un-Leibnizian way of speaking—the force belongs to one substance rather than to the others, and this is the "cause" of the change.

In the correspondence with Clarke it seems that the Newtonian and Leibnizian cosmogonies could scarcely be farther apart. Yet the differences can be exaggerated, for what separates the two is mainly metaphysical differences for which the physics is largely irrelevant. This is particularly evident over the concept of force. For Newton the existence of certain kinds of force directly implies the existence of certain kinds of motions, and these in turn can only be accounted for by postulating an extraordinary frame of reference. For Leibniz, on the other hand, all phenomenal motion is relational; yet it is not all metaphysically equivalent. Leibniz is prepared to concede that there is a difference between "absolute true motion" and mere "relative change of situation". A body is truly in motion when the immediate cause of the change is in that body rather than another. And although it may be impossible for us to know in which body the cause of the change inheres, it is certain for Leibniz that such facts are known by God.

Thus for both Leibniz and Newton forces—qua causes of phenomena—are ultimate metaphysical realities. Where they disagree profoundly is that for Newton an explanation of these causes may be forever opaque to the intellect, and the discussion of such explanations must remain hypothetical only, and beyond the task of a mathematical science of nature. For Leibniz, the principles upon which the whole world of experience is based can be apprehended by human reason.

These differences—and similarities—can be highlighted by putting the respective positions into a Kantian mode of speech. For Newton, the world of phenomena is all that we can know in science; the ultimate entities are real enough—like absolute space and time—but they function as "regulative ideas". They complete the metaphysical picture, but are not themselves objects of possible experience. The sciences can postulate them as metaphysically necessary, yet at the same time deny them operational relevance. Leibniz uses a different kind of argument entirely. For Leibniz, *this* is the way the phenomenal world is. Given this as *quid facti*, the laws and principles that produced it and made it possible must then be of such-and-such a form. If Newton in his positivistic moments was an empirical realist, we can perhaps see how, by means of a radical approach to the very conceptual problems that separated Newton from Leibniz, viz. the nature space and time, Kant was led to the metaphysical position of transcendental idealism as the only solution to such conflicting claims by means of a re-shuffling of the philosophical cards.

The Leibnizian relation to Kant can be made more precise. Neither space nor time are real; they are determinations or systems of relations in which, and by means of which, the real world is given to us. Neither

Leibniz nor Kant would consider them abstractions, in the sense that space and time are not concepts derived from already given spatio-temporal elements. As Leibniz writes: "...abstraction is not an error, provided we know that what we are ignoring is really there". Also, in his critique of absolute motion, Leibniz appeals to his proto-verificationist principle; what cannot be observed is in principle no object of possible experience. Leibniz's use of such a positivist line often seems limited to the problems of space, time, and motion. He does not feel the necessity to apply any such rigorous criterion of significance to those elements of his system considered metaphysically essential, i.e. the monads themselves.

The connection and mutual relevance of physics and metaphysics was never doubted by Leibniz. The phenomenal world, including the world of physics, is a confused expression of the order and harmony which exists on the metaphysical level. Everything is ultimately explicable in terms of the monads, their states, and phenomenal perceptions of these states. The vectorial nature of velocity, for example, is a confused representation of the goal-directedness of monadic appetition. It is because this process—infinitely complex, continuous, and unique for each monad—is the metaphysical explanation for all phenomenal occurrences, that Leibniz considered the framework of absolute space and time to be unnecessary. Newton, having idealised bodies into mass-points, indistinguishable from one another, needed the framework of absolute space to differentiate these mathematical bodies: they needed "room to be different in". Just as it is Leibniz's God who, acting perfectly and logically, creates an infinity of substances which He at least can identify individually: so it is Newton's God who individuates the idealisations of Newtonian physics by this unique and infinite system of differentiated positions in space and time. But an infinite space and time must be dependent on God, like everything else; so Newton's God is also present everywhere and at all times. "He is not eternity and infinity, but eternal and infinite: He is not duration or space, but He endures and is present". [In H.G.Alexander, p.167.]

I suggested above that the motives for the Clarke-Leibniz correspondence were as much theological and metaphysical as scientific, the similarities in terms of dynamics and the description of nature being frequently overlooked in favour of the polemics. Theologians such as Raphson used Newton in support of purely theological doctrines, which probably has as much to do with the attacks made on Newton by Leibniz (and also Berkeley), as did the technical and difficult problems of pure natural science. This does not obscure or diminish the value of the debate, which does not end here, but is transformed subsequently by Kant.

Chapter Two

KANT'S THEORY OF SPACE AND TIME

2.1 Introduction

The problems encountered in Chapter One really arise from the same source, namely, transcendent metaphysics. In spite of Newton's attempts to demonstrate the reality of absolute space and time, their postulation as realities remains always a metaphysical article of faith. Newton's arguments, in their analogical form, are suggestive but not conclusive. Leibniz's response was ingenious, but lacks ultimate conviction because his conclusions are drawn from premises which do not have the self-evidence for us that he claimed for them. The debate between Newton and Leibniz is often pursued at a high level of abstraction; in both cases we are required to move beyond and, in a sense, "behind" appearances, in order to glimpse transcendental realities—in one case, absolute space and time; in the other, fundamental principles from which all knowledge of reality is supposedly deduced.

There are no theories of space and time that are isolated from wider philosophical concerns. Newton, for instance, was both an empirical scientist and a Christian; his attempt to construct a theory of space and time compatible with each of these aspects of his character inevitably produced tensions, some of which have been indicated above. Leibniz, carrying a rationalist banner, was driven to find indubitable metaphysical principles from which everything that may be known could be deduced. We have seen that by limiting the ultimate realities to internal states of monads, Leibniz had great difficulty in accounting for space and spatial relations which seem to demand, if anything does, some "external" structure. We shall see in what follows that in important respects Kant abandons neither of these positions in their entirety. In particular, Kant's problem can often be traced to these earlier debates even when his solutions diverge so radically as to sometimes disguise the historical links. Clearly the absolute and all-embracing framework of the space and time of Newton has certain methodological advantages; and Leibniz's insistence on principles of significance has great therapeutic force as a prophylactic against baseless metaphysical speculation against which Kant was to rail consistently. And if Leibniz did not always adhere to their strictures, then Kant will certainly remind him just when and where Leibniz's own principles have deserted him.

For Kant the radical sceptical attack on knowledge held little interest in itself. For him the real problem was not do we know such-and-such, but *how* do we know. Newtonian physics and Euclidean geometry were, for Kant, examples of genuine knowledge. How were they possible? What was the structure of experience like, and what was the relationship between the objective world and the mind which enabled this knowledge to arise? And why had metaphysics never achieved the same level of progress and conviction?

It was to these questions that Kant addressed himself; and it was the discovery of an entirely different source of real knowledge that led to the "critical" philosophy. It was the recognition of the difference in kind between mathematical and philosophical knowledge that first stimulated Kant's imagination. The nature of the propositions in these two domains, their source, their method of validation, leads directly to an entirely new and provocative theory of space and time. In what follows I will consider Kant's theory of space and time in the context of immanent metaphysics; this requires an exposition of his theory of concepts and propositions; a discussion of the Transcendental Aesthetic—Kant's reply to Newton and Leibniz, as well as a powerful new theory of his own; the transcendental idealist solution to the problems raised by rationalism and empiricism; and, finally, the perennially interesting problem of incongruent counterparts and its relevance to the problem of the nature of space. I will suggest that from all of this emerges a coherent Kantian model of space, time, and mathematics.

2.2. Concepts and Definitions

In the introduction to the *First Critique* Kant writes that an analytic judgement is one in which the connection between subject and predicate is "thought through identity", which suggests that the truth of a proposition can be inferred from the fact that it has a self-contradictory negation. The predicate adds nothing to what is already "contained in" the concept of the subject. In Kant's words it is an "explicative" judgement, (*Erläuterungsurteile*). An analytic proposition, though it may help us understand the subject concept better, perhaps by making explicit connections previously thought only confusedly, cannot actually increase our knowledge. Such judgements give us clarity rather than extend our knowledge as such. Such a concept of analyticity is not unproblematic, involving as it does two criteria for differentiating analytic from other kinds of judgements. The first criterion is conceptual and depends on the somewhat loose notion of a predicate being "contained in" a subject concept; the second criterion is purely logical, reducing as it does to the application of the principle of contradiction. (The "containment" metaphor clearly reminds us of Leibniz.) These two aspects of the notion of analyticity are not only related so far as Kant is concerned, but frequently confused by him.

We can see how Kant thinks they are related by turning to a source outside the critical philosophy. In his *Logic*, Kant considers the following to be one of the "...universal merely formal or logical criteria of truth, viz. the Principle of Contradiction and of Identity, by which the intrinsic possibility of a cognition is determined for problematical judgements". [Kant, *Logic*, p.43.] In the same work he gives additional clarification of the idea of analyticity, this time in connection with a discussion of attributes. It is by means of the latter that cognitions are possible. An attribute, Kant says, is "...a partial conception so far as it is considered as a ground of cognition of the whole conception". [*Logic*, p.48.] All concepts are attributes, and all thinking is conception by means of attributes. These may be either "conceptions in themselves", when they can serve as the subject term in a judgement; or "partial concepts", which are "grounds" of cognition of the subject concept.

It is not only complete judgements that can be divided into analytic and synthetic modes, but attributes also. Indeed, the notion of a predicate being "contained in" the subject concept is clearer if one considers the analytic and synthetic nature of attributes by means of which complete judgements are possible. An analytic attribute is a partial conception of the actual concept which is already thought in it; while a synthetic attribute is a partial conception of the "merely possible" total concept, the latter being a synthesis of parts. Briefly, analysis consists of making a concept distinct, while synthesis makes a distinct concept. It seems clear that for Kant the purely logical aspect of analysis is inseparable from the analysis of concepts. (It should be emphasised that we are concerned, for the moment, only with conceptual analysis in philosophy: very different considerations apply in the case of mathematical definitions, and I will discuss this below.) This helps us to understand how and why Kant thinks he is entitled to assert in the *First Critique* that analytic judgements are always adequately known to be true in accordance with the principle of contradiction:

> The proposition that no predicate contradictory of a thing can belong to it, is entitled the principle of contradiction...The reverse of that which as concept is contained and is thought in the knowledge of the object, is always rightly denied. But since the opposite of the concept would contradict the object, the concept itself must necessarily be affirmed of it. [B191]

Thus Kant affirms the principle of contradiction as a sufficient criterion of truth—but only in respect of analytic judgements.

In analysis, the whole precedes the parts, and so no new knowledge of objects is possible through this method. To extend our knowledge we

CHAPTER TWO – KANT'S THEORY OF SPACE AND TIME

must employ synthetic judgements. Such judgements extend our knowledge by predicating something of a concept which would not have been possible by mere analysis. Kant calls these judgements "ampliative" (*Erweiterungsurteile*). The examples he gives are, for analytic judgements, "All bodies are extended"; and for synthetic judgements, "All bodies are heavy". The latter—taken in the context of Newtonian physics—is synthetic because it is sometimes false, for example when a body is isolated from other large masses and has no "weight". All judgements of experience are synthetic. It would be absurd, according to Kant, to base an analytic judgement on experience, although some analytic propositions are instantiated in intuition; for example, *a* is *a*.[1] This is a consequence of Kant's definition of an analytic judgement, that is, one in which the negation of a proposition leads to contradiction. It also emphasises the logical as opposed to the conceptual aspect of analyticity.

This is not, it must be said, a particularly satisfactory notion of what an analytic judgement is, since we know that a proposition is self-contradictory only because we already accept that the original is analytic. That is, we would not know or accept that not-P is self-contradictory unless P was analytic. However, taking these examples as typical, we could say that given a conceptual interpretation of analyticity, there seems no reason why empirical concepts should not be the foundation for analytic judgements. The concept of "body" is empirical, even if idealised. The fact that extension is connected with "body" necessarily, while weight is connected only contingently seems, on the face of it, to reduce to a question of an exhaustive definition of the concept of body.[2] If we chose to define an empirical concept differently, there is no reason why we should not make analytic judgements founded upon it. But caution is required here, if Kant is not to be misrepresented. He does, nonetheless, seem to suggest that an analytic judgement—or rather, the means by

---

[1] It is difficult to be completely sympathetic with this point. Although obviously true, it would have to be a very odd kind of experience that might lead someone to affirm it.
[2] An important argument for this idea that empirical judgements might justify analytic propositions can be found in Kripke, (1972).

which we know it—and a definition, are one and the same. They differ in that a definition must exhaust the concept, while a single analytic proposition is only a partial conception of such an exhaustive definition. "To define...", says Kant, "...really only means to present the complete original concept of a thing within the limits of its concept." [B756]. This seems to mean no more than that an exhaustive definition of a concept would be a list of its predicates, and that one of these predicates would be, in this example, the analytic judgement "all bodies are extended", or "this body is extended". [Cf. A241.]

Analytic judgements are, says Kant, stages in the clarification of concepts, and definitions are the end-product of this process of thought and not the "static circle of tautology".[3] My suggestion above concerning analytic judgements based on empirical concepts here seems to be denied by Kant, who says that since definition is "...no thing more than the determining of the word...", then the only concepts that can properly admit of definition—that is, can be exhaustively defined—are those which are arbitrarily invented or constructed. "There remain therefore, no concepts which allow of definition, except only those which contain an arbitrary synthesis that admits of a priori construction. Consequently, mathematics is the only science that has definitions." [B757]

In the Discipline of Pure Reason, Kant argues that mathematics owes its exactness to its legitimate possession of definitions, axioms, and demonstrations, none of which is available to the philosopher. I shall be making a good deal of this distinction in later sections, so it will be useful to look at what Kant has to say about definitions at this time. As we have seen, he says that to define "...really only means to present the complete, original concept of a thing within the limits of its concept".[4] Kant then goes on to say that if this is the standard for "definition", then empirical

---

[3] See also B759a, where Kant admits the usefulness of incompletely defined concepts in philosophy.
[4] Kemp Smith's translation is somewhat uncomfortable here. The original reads as follows: "Definieren soll, wie es der Ausdruck selbst gibt, eigentlich nur so viel bedeuten, als, den ausführlichen Begriff eines Dinges innerhalb seiner Grenzen ursprunglich darstellen...". See Kant, *Kritik der Reinen Vernunft*, 2, (ed. W.Weischedel), B755.

concepts may not be defined at all but only made "explicit". Empirical concepts are "open": they cannot be limited arbitrarily in their intension. Neither may pure concepts be defined in this way, since we cannot know that our representation as given is not "confused", until we have found an intuition adequate to it. The best that we may hope for in this regard is the probability of completeness, not its certainty. Since Kant has ruled out both pure and empirical concepts from within the range of definition, all that remains is that kind of concept that has been "invented", viz. mathematical concepts.

However, the fact that a concept can be arbitrarily invented, as in mathematics, does not mean that the object of the concept has thereby become a "true" object. [A729/B757]. Kant says that such arbitrary invention should be confined to mathematics, since mathematics is able to construct its objects in pure intuition. An "invented" concept that is not mathematical may depend on empirical conditions: we can thus have no assurance of the object's possibility. Such a concept, says Kant, would be better described as "...a declaration of my project than as a definition of an object." Kant concludes from this that there are no concepts that can be defined, "...except only those which contain an arbitrary synthesis that admits of a priori construction". [A729/B757].

It is not clear that Kant has prepared us for this point—that is, from the arbitrariness of a definition to the possibility of a priori construction. He is clearly assuming that this relates to mathematical objects, and he goes on to distinguish definitions from demonstrations. As in the case of the former, the only proper domain for the latter is mathematics. For Kant, demonstrations are intuitive proofs grounded in constructions.

This characteristic view of Kant's that definitions are the beginning of mathematical reasoning but the achievement, the end product, of philosophy, immediately raises problems. Since the proposition "All bodies are extended" is not mathematical, this cannot be a definition of "body": no concept given a priori can be defined. How then do we decide that it is analytic? Kant seems to be suggesting that it is part of the meaning of "body" that the predicated "extended" should be applicable

59

to it. But if it is analytic because of the meaning of the word, it does seem to be a matter of definition.

There is much in this analysis reminiscent of Leibniz's notion of finitely and infinitely analytic propositions, which Kant in fact rejects. If we consider contingent subject-predicate propositions by analogy with analytic propositions, it is perhaps easier to understand how Leibniz came to his "predicate in notion" principle, as well as the consequences this has for his system. If we say that the predicate of an analytic proposition is contained in the subject concept, we assert that S is P when we really already know that P is "in" S. Suppose that we know, by definition of S, that S is exhaustively defined by predicates $p1, p2, p3...pn$; then since S is really a shorthand way of referring to this finite conjunction of predicates, the proposition that S is P is analytic. If we then extend this to contingent propositions, it is easy to see how Leibniz came to the idea of "infinite analyticity". The predicate P is still contained in the subject *concept*, but since the list of predicates is infinite, only an infinite mind can see that it is necessarily true that S is P. However, this will work only so long as we accept that concepts—of all types—can in principle be exhaustively defined. It was Kant's achievement to have seen that such exhaustive definition belongs properly only to concepts that have been arbitrarily constructed; with such concepts it is clear that true propositions concerning them must be analytic. But by trying to avoid the implications of Leibniz's confusion of "logical subjects", and the *concepts* of these subjects, Kant himself conflated the logical and conceptual aspects of propositional analysis.[5] By restricting the meaning of analyticity to predicates contained in subject concepts, and limiting definitions to mathematics, Kant has given his own examples of synthetic judgements an air of arbitrariness; while the analytic judgements appear as "conceptual necessities" or mere linguistic conventions.

Although there are profound and unresolved problems raised by Kant's formulation of analyticity, he seems not to be unduly concerned

---

[5] For an important discussion of the distinction in Kant between two senses of "object"—viz. "*Gegenstand*" and "*Objekt*", see Henry Allison, (1983), p.27-28.

by them. In any case, most philosophers since Kant who disagree about candidates for analytic propositions generally agree on characterisations for synthetic propositions. The latter are genuinely informative, and they function often as "hypotheses" that try to express something true of a concept which is not, and could not be, exhaustively defined. With all non-mathematical concepts we aspire, by means of philosophical analysis. to definitions. From "expositions" we gain philosophical knowledge: from definitions, we gain mathematical clarity.

The tenor of Kant's argument moves against the rationalist ideal of philosophical knowledge through concepts alone. It is perhaps one of Kant's greatest achievements to have discovered that the dichotomy of logic and experience does not exhaust knowledge. The reason both Hume and Leibniz believed in the exhaustiveness of this dichotomy was that they confounded analyticity with a priori necessity. Kant insists on their independence, arguing that the former is essentially logical, the latter essentially epistemological.[6]

It is on these foundations that Kant develops his "anti-logicist" programme. An analytic proposition is nonetheless always a priori, since the way in which it is justified coincides with the way in which we recognise it as analytic.[7] Taking the containment metaphor again, we can see by analysis that the predicate belongs to the complete definition of the subject concept. The appeal to the logical criterion tells us that an analytic judgement cannot be negated without contradiction. For Leibniz, at least, the former criterion was reducible to the latter. Because analytic judgements are such by virtue of the predicate being part of the concept of the subject, it is itself an analytic truth that no analytic judgement is a posteriori. It cannot be objected that we call a judgement analytic because it is a priori, since there would then be no judgements which were

---

[6] The distinction can be broadly summarised as that between propositional structure and propositional justification. For an analysis of this idea, discussed within the context of justifying the notion of the synthetic a priori, see the seminal paper by N.R.Hanson, "The Very Idea of a Synthetic A Priori"; in Sumner and Woods, p.65-70.
[7] It must be acknowledged that Kant comes perilously close to confusing propositional structure and propositional justification himself.

synthetic and yet a priori: and there is at least one such judgement—namely, the judgement that all knowledge is based on either logic or experience. This disjunction must be, as a principle, considered as necessary (a priori) and non-analytic, since it is intended to be—and if it is either true of false, then it is—genuinely informative. Analytic judgements are in effect in need of no additional justification. They are justified by virtue of their logical form, so to say that a judgement which is analytic is also a priori, is to assert an analytic judgement.

2.3 Kant's Anti-Logicist Programme

The whole of the First Critique is an attempt to answer the question: How are synthetic judgements, known a priori, possible? The account of analyticity just given has shown that for Kant analytic judgements cannot, as such, add to our knowledge in any important fashion. Kant does not ask whether there are any synthetic propositions known a priori, but how they are possible. He takes it that all propositions of any truly "scientific" metaphysics—that is, one which yields real knowledge—are of this type. The question as to how they are possible is answered by examining their application in the realm of mathematics.

According to Kant, all judgements of mathematics are synthetic. There are "fundamental" propositions, he concedes, that are "presupposed by geometers" that are analytic, based upon the principle of contradiction as grounds for their truth: for example, $a=a$, or $\{a+b\}>a$; but such propositions are tools of analysis and are used only as links in the chain of method, not as principles. [B16; See also Poincaré, p.35.] This fact, says Kant, has misled philosophers into assuming that all of the propositions of, say, geometry, can be validated by means of the purely logical criterion of the principle of contradiction. It is true that synthetic propositions may be discovered through this principle, but only when other synthetic propositions are presupposed. What Kant is suggesting is that such logical principles allow the transmission of one informative judgement to another; but the truth of the conclusions of such a transmission are not reducible to the vehicle of transmission itself.

Many philosophers find difficulty in accepting even the possibility of synthetic judgements known a priori. One source of confusion has been the tendency to blur the lines of demarcation between the analytic and the synthetic on one side, and the a priori and a posteriori on the other. Indeed, some philosophers regard the terms "analytic" and "a priori" as synonymous. If one takes this line, the "problem" disappears, since the term "synthetic a priori" becomes self-contradictory. It is certainly true that Kant does not feel that the a priori/a posteriori distinction is in need of elaborate justification, since this division of validatory source is taken over without essential change from the rationalist tradition that Kant inherited. (We might also note that Kant claimed to have invented the analytic/synthetic terminology: in a letter to Eberhard, Kant felt able to say that therefore "analytic" had to mean what he—Kant—said it meant!) The concern of the First Critique is "Transcendental" philosophy, that is, with the a priori conditions which make knowledge possible. If we can be clear about the meaning of a priori necessity, it is surely in this context. To say that a proposition is known a priori and yet is synthetic, is to include it in any philosophy that concerns itself with how a certain kind of knowledge is possible.

The idea of a synthetic proposition, known a priori, is, as such, quite consistent. There is nothing intrinsically absurd about an "ampliative non-empirical proposition"; that is, a proposition whose truth is not derived from experience, but which nevertheless adds to our knowledge. And for Kant, this is precisely what the propositions of mathematics are like. He argues that there is nothing in the concept of the union of 7 and 5 which allows us to obtain the "concept" of the number 12. To reach the latter we must enlist the aid of intuition, which is the only means by which we may go beyond one set of concepts to another synthetically.

It has to be said that Kant adds to the exegetical problems here by his shifting use of "concept" language, for he writes of the "union" of 7 and 5, (..."die Vereinigung beider Zahlen in eine einzige..."); but if we look at a later statement of this example from the First Critique (in the Axioms of Intuition), we find that he is rather more revealing:

> That 7+5 is equal to 12 is not an analytic proposition. For neither in the representation [*Vorstellung*] of 7, nor in that of 5, nor in the representation of the combination of both, do I think the number 12. (That I must do so in the addition of the two numbers is not to the point, since in the analytic proposition the question is only whether I actually think the predicate in the representation of the subject. [B205]

Kant may be correct in asserting that the mere placing together of 7 and 5 does not lead us, without recourse to intuition, to the concept of 12. However, the formula "7+5=12" seems to be a subject-predicate proposition only by courtesy. If we re-cast it as "the sum of 7 and 5 is 12", we might, at a pinch, take the term "the sum of 7 and 5" as the subject of the proposition. Even so, it is difficult to take it seriously as a subject *concept*.

Kant seems to be operating with two criteria for analyticity here which though complementary, are not equivalent. For example, when discussing the proposition "all bodies are extended" Kant appeals to the principle of contradiction; it would be contradictory to assert that "some bodies are not extended", since extension is part of what we think in the concept of body. The proposition "all bodies are heavy" is synthetic because heaviness is not part of the concept of body. The synthetic element in mathematical propositions must be considered a consequence of their constructive nature. We might argue that Kant is prima facie correct in asserting that "12" is not included in the concept of the union of 7 and 5; and that in order to establish the truth of the proposition 7+5=12 we must "exhibit" the concepts in intuition—that is, we must perform the operation of counting. But are we then not compelled to say that we cannot know a priori that we will in fact reach 12 after performing this calculation? Is not the proposition thereby seen to be synthetic and a posteriori?

Kant answer is that this is of course mistaken. The construction is intuitive, and the intuition is pure, i.e. non-empirical. Let us agree with Kant and say that 7+5=12 is synthetic, because we must go beyond the concept of the union of 7 and 5 in order to reach 12; and we accept this because in order to validate such a proposition we resort to an intuitive construction. In other words we reject, with Kant, the claims made by empiricists and rationalists alike that mathematical propositions are reducible to logic. We accept that something more is needed to validate such propositions than appeals to the law of contradiction.

Have we now demonstrated that all mathematical propositions are synthetic? Consider, not 7+5=12, but 12=7+5. Is it not possible to suggest that it *is* part of what we mean by "12" that it analytically contain the sum of 7 and 5? (We can only get away with this at all thanks to the ambiguity of the treatment of subject-predicate propositions in Kant, in relation to mathematical formulae which are really statements of equivalence. If "12" is the subject, and "7+5" the predicate, then the proposition 12=7+5 is not equivalent to 7+5=12). So although it is a question of a priori construction to validate 7+5=12, once this has been validated, is it not analytically true that 12=7+5? Kant's answer completes his theory of number (which we shall return to below in a different context) relating the process of a priori construction to intuition, the latter taken not in its formal and original sense as a "particular" rather than a general concept, but in its full transcendental idealist meaning as related to sensibility and its a priori forms, space and time.

For Kant the employment of numbers is quite general. They are and must be independent of the synthesis of their generation. This is based upon the possibility of apprehending the manifold both forward and backward. A synthesis once performed does not need to be performed again and again: it is valid once for all, and the symbolic construction, i.e. the numerals etc., by which it can be replaced, represents the general application of a symbol to the successive synthesis.[8] It is misleading for

---

[8] There is a similarity between this and Philip Kitcher's argument justifying an empiricist philosophy of mathematics. Kitcher says that "...we collect prior operations. But in

Kant to refer to the "concept" of 7 and the "concept" of 5 (as he does in the *Introduction* to the First Critique), since he does not regard each number as in some sense a general name for a class. If one defines a number like this (as in Frege, for instance) it would be plausible to say that there was a separate concept for each number. Taking the "concept" of 7 and the "concept" of 5 perhaps it is impossible to arrive at the "concept" of 12 without the aid of intuition, i.e., without making the calculation. But if instead we consider the general concept of number, we might say that the union of any two numbers involves, or contains, the product. This would require some functional meaning as part of the concept of number, some rule of generation by means of which we can construct any number, any particular member of the series, whatever. As individual elements numbers would take their essential meaning from the whole series of which they are a part.

In the passage from Kant cited above, Kant—who as we have seen frequently uses the term "concept" (*Begriff*) somewhat loosely—refers to numbers as "representations", *Vorstellungen;* this has the merit of reminding us that mathematical symbols are not discursive general concepts, but particulars constructed in intuition. The active, functional characteristic of number is related to pure intuition by means of its relationship to time. Kant expresses this—though the idea is little developed—through allusions to time as being related to arithmetic as space is related to geometry:

> Hence pure mathematics deals with space in geometry, and time in pure mechanics. In addition to these concepts there is a certain concept which in itself indeed is intellectual, but whose actuation in the concrete requires the assisting notions of time and space (by successively adding a number of things and setting them simultaneously beside one another). [P.C. p.62]

---

performing these operations we use symbols. To collect is to achieve a certain type of representation and, when we perform higher-order collectings, representations achieved in previous collecting may be used as materials out of which a new representation is generated". Kitcher also recognises the link with Kant's theory of algebra explicitly, in spite of the general anti-Kantian tone of his book. See Kitcher, p.129ff.

This direct relation of time to number is expressed in various ways by Kant, and appears again in the difficult Schematism chapter. Number is there said to be "the pure schema of magnitude"; it is a representation "...which comprises the successive addition of homogeneous units. Number is therefore simply the unity of the synthesis of the manifold of a homogeneous intuition in general, a unity due to my generating time itself in the apprehension of the intuition". [A143]. I will postpone a fuller discussion of these cryptic and important sayings, but for the present we can say that for Kant, the symbolic representation of a particular position in the ordered sequence generated by time, is number:

If, in counting, I forget that the units, which now hover before me, have been added to one another in succession, I should never know that a total is being produced through this successive addition of unit to unit, and so would remain ignorant of the number. For the concept of number is nothing but the consciousness of this unity of synthesis. [A143].

Kant claims to have discovered judgements that are informative—synthetic—and yet independent of sense-experience, and are thus a priori. The predicate of any such judgement is related to the subject concept by means of an intuition that goes beyond the subject concept itself for its validation.

2.4 Transcendental Aesthetic

Kant has argued that the example of pure mathematics is evidence for the existence of propositions that are synthetic and yet a priori. Mathematics adds to our knowledge, and is not reducible to logic, which gives only formal criteria for the truth of analytic judgements. The Aesthetic is part of Kant's attempt to show how such judgements are possible.

Although the Transcendental Aesthetic offers a theory of space and time, this is not its sole concern. The theory of space and time is an answer to a question, an attempted solution to a profound metaphysical problem. Long before the critical philosophy was a reality Kant had recognised a fundamental difference between the methods of philosophy and mathematics. The success of the latter in achieving not just logical certainty but also truth in its applicability to the world, deeply impressed Kant as it had impressed every thinker before him. Kant summarised his view of the difference by saying that philosophical knowledge is knowledge gained by reason from concepts; mathematical knowledge is knowledge gained by reason from the construction of concepts in pure intuition. [B741].

In enquiring into the nature of our representation of space and time, Kant asks that they must be like in order that judgements regarding them can be verified a priori. The Aesthetic is also, we should remember, a discussion with Newton and Leibniz, both of whose theories are criticised essentially because neither can provide for the possibility of synthetic a priori judgements as they are found in science and mathematics. So the Aesthetic has the role of explaining how such knowledge is possible by offering a theory of space avoiding the mistakes of Kant's predecessors. This indicates a close relationship between his theory of space and his conception of geometry; a relationship expressed through the metaphysical and transcendental expositions of space. Some critics suggest that since each aspect—the theory of space, and the nature of geometry—is for Kant the condition for the possibility of the other, then if the theory of geometry falls, then so too must the theory of space. I will argue that the two theories are sufficiently independent of one another to permit one to be held without an equal commitment to the other.

Kant's describes a metaphysical exposition is the "...clear, though not necessarily exhaustive, representation of that which belongs to a concept: the exposition is metaphysical when it contains that which exhibits the concept as given a priori". Such an exposition thus makes implicit assumptions explicit. The propositions that Kant is aiming to demonstrate thereby are as follows: that our representation of space is not derived from experience; that it is an a priori representation; that it is an intuition and not a general concept; and that it is an infinite given magnitude.

The first argument is an attempt to show that our idea of space could not be derived from empirical objects and their relations, *pace* Leibniz, since for the possibility of such objects being objects of experience "...the representation of space must be presupposed". [B38]. External perceptions presuppose the concept of space and do not create it. Kant argues that since we cannot represent to ourselves objects without space, yet we can represent to ourselves space without objects, the former must be the condition for the possibility of the latter. However, from this argument (which is in essence the same as the first), Kant reaches a more radical conclusion, viz. that space is a necessary representation a priori. The usual translation of Kant's "...eine notwendige Vorstellung a priori..." as "necessary a priori representation", is liable to mislead; the translation suggests the idea of an innate concept, which the original does not, at least much less so. The point is that the suggestion of "innatism" was strongly resisted by Kant, as this passage from the *Inaugural Dissertation* testifies:

> The question arises for everyone as though of its own accord whether each of two concepts [space and time] is born with us or acquired...The former must not be so lightly admitted since it paves the road for a philosophy of the lazy, a philosophy which by citing the first cause declares any further search vain. But truly *each of the concepts* without any doubt *has been acquired*, not by abstraction from the sensing objects indeed... (for sensation gives the matter and not the form of human cognition), but from the very action of the mind, an action coordinating the mind's *sensa* according to perpetual laws, and each of the concepts is like an immutable diagram and so is to be cognised intuitively. For sensations excite this act of the mind but do not influence the intuition. [P.C. p.73].

This passage is of great interest. The remark to the effect that space and time are like "immutable diagrams" contains pre-echoes of the later doctrine of schematism. What is more to our immediate point is that while Kant firmly repudiates the idea that we have innate knowledge of space and time, prior to experience, the passage also demonstrates how Kant's own preoccupations colour his results. The sentence "...sensation gives the matter not the form of human cognition" really assumes, without argument, that the form could not be given along with, as it were, the matter. In the First Critique Kant does admit that the separation of the form from the content of experience is admissible only for the purposes of clarification—a metaphysical exposition—of which the Aesthetic is an example. [B35]. And reference to the "concept" of space shows that at this time the discovery of space and time's intuitive nature had not been fully recognised.

Kant argues that since we can represent to ourselves only one space, then space cannot be a "discursive" concept. This is a view expressed by Kant in several places and amounts to the idea that whereas concepts contain their instances *under* themselves, intuitions contain their instances *in* themselves. That is, while the concept "red" is not itself red, the intuition of space is itself a space. Parts of space are merely limitations of this one all-embracing space. These parts cannot be constituents out of which the one all-embracing space is composed. The parts cannot precede the whole as is the case with concepts, where we abstract the relevant characteristic from the various objects and form the concept. (This is one way, at least, in which we might achieve this.) There is also the well-known "contradiction" in Kant between the space and time of the Aesthetic, where the whole is said to precede the parts; and the space and time of the Axioms of Intuition, where space (at least) is said to be known through "...successive synthesis of part to part". [B204]. It is sometimes argued that the Axioms of Intuition are a trivial consequence of the conclusions of the Aesthetic, or otherwise that they are inconsistent with those conclusions. [See Brittan, p.90]. Generally, I have kept Kant's ideas on space and time distinct, not by emphasising any supposed differences of treatment they are given in the Aesthetic and the Analytic respectively—on this, Kant has good methodological grounds for treating them in different ways: but by keeping in mind the distinction between spatiality—the "amorphous continuum"—and determinate space. In the Axioms, Kant has the latter in mind: "...the mathematics of space (*Ausdehnung*)—geometry—is based upon this successive synthesis of the productive imagination in the generation of figures". The manifold of space is really a fiction and depends on limitations. For instance, a part of space is limited by the space outside it: indeed, it is not really a part at all, since the parts constitute the whole only in an ideal manner. [See Nagel, Chapter 4]. The parts have reality, as parts, only because the whole exists prior to them, whereas with a concept the parts truly make up the whole—the whole is composed of parts and it is the whole that has merely ideal existence.

There is an interesting distinction in Kant's expression of these ideas in the Aesthetic and the earlier *Inaugural Dissertation,* which contains embryonically almost all of Kant's mature critical views on space and time. In the latter Kant is much more explicit about precisely what kinds of entities space and time are; and the difficulties with this view are correspondingly more evident. He writes there of the *concept* of space, and the *idea* of time—almost as if at this stage he wasn't really sure what they should be called. In the mature philosophy there is no doubt that they are to be regarded as forms of intuition (or sometimes as "pure intuitions" *simpliciter*), and are not concepts. The view sometimes attributed to Kant that space is somehow "injected" into a non-spatial world is less plausible if in each instance where we refer to space as such, we talk instead of the "idea" of space. There is a difference between the assertion, let us say, that *space* is an a priori particular, and that our *idea* of space is such. On a psychologistic reading of Kant, we might ask how it is possible for there to be one infinite given space for each thinking and perceiving mind. On such an interpretation we are reduced to two alternatives: we may either appeal to a pre-established harmony between the representations of different finite minds, in order to guarantee the move from subjective to inter-subjective space; or, treating the argument as a piece of faculty psychology, appeal to empirical rather than transcendental considerations. We may assume that Kant would have rejected both of these.

Even if we accept that Kant has shown that space is a "singular" representation—an intuition in the more limited sense that implies that it is a particular rather than a general concept—we may still ask whether Kant has shown that this intuition must be pure, not empirical; without, that is, an implicit appeal beyond the metaphysical exposition to the transcendental exposition, which would take us directly into his entire theory of mathematical construction. His response might be that he has already shown the representation of space to be a priori, so the intuition must be pure. But there is no doubt that Kant had great difficulty in keeping the metaphysical exposition—the *quid facti*—separate from the transcendental exposition, the *quid juris*. Not altogether surprisingly, the two are completely mixed up in the *Inaugural Dissertation*, just as they are in the first edition of the *Critique of Pure Reason*. Consider the following passage, included in the first edition, but omitted in the second;

> The apodeictic certainty of all geometrical propositions and the possibility of their a priori construction, is grounded in this a priori necessity of space. Were this representation of space a concept acquired a posteriori...the first principles of mathematical determination would be nothing but perceptions. [A24].

The fourth argument of the metaphysical exposition is perhaps the most difficult for the sympathetic reader to grasp. "Space", says Kant, "...is represented as an infinite given magnitude". Let us recall at this point what Newton said about infinity. Newton says that "...'end', (*finis*) is a word negative to sense, and thus 'infinity' as it is the negation of a negation, will be a word positive in the highest degree with respect to our perception and comprehension, though it seems grammatically negative". Now if we take this in conjunction with the view about the parts of space being ideal limitations of an all-embracing space we can, I think, begin to make sense of what Kant says. Just as for Newton, Kant holds that the infinite contains the ground for the finite, as the latter must always be known as a boundary that necessarily points beyond itself. It is, says Kant, only when infinite space and time are given that definite magnitudes can be assigned to limitations in space and time. Like Leibniz, Kant regards points and instants as limits, not parts, of space and time. What Kant seems to have in mind here is that all measurement requires the recognition of a definite boundary of a quantity which must contain the definite as part of itself. The idea was formulated by Kant in his "Concerning the ultimate foundation of the differentiation of regions in space", (of which much more below), where he says that the parts of space presuppose a region according to which they are ordered; and also in the *Inaugural Dissertation*, where he states that "...it is the infinite which contains the ground of each part that can be thought and finally of the simple or rather the boundary. For it is only when both infinite space and time are given that any definite space and time are assignable by limiting". [P.C. p.72].

Kant is here making one of his moves from what we might call "anthropological" facts, to transcendental necessities. He cannot make intelligible the idea of a temporal or spatial boundary which did not "point beyond" itself; it follows from this that space and time must be homogeneous and continuous, for there could be no limit to space and time in either direction of magnitude. Indeed, the very idea of *a* space and *a* time being definite, is characterised by something outside it, which limits it. From this, Kant moves to the assertion of the necessity of the *postulate* of infinite space, which can in principle contain any and every space. Each of these finite spaces can in turn contain other spaces, and this guarantees that space is continuous. Space and time are thus *quanta continua*; "Space consists solely of spaces, time solely of times". [A169]. Positions presuppose the intuitions which they limit, and from this system of non-spatial positions space cannot be constructed, *pace* Leibniz.

It is clear that when Kant asserts that the representation of space is an "infinite given magnitude", this must be understood not as a claim about the extension of physical space, but as a claim about what meaning is to be given to ideas of infinity and continuity from within transcendental philosophy. When Kant writes of intuition in the Aesthetic, he is making the connection with sensibility directly. If we restrict the notion of intuition to its meaning as "particular" we could make the distinction—essential to Kant's conception of mathematics—between concepts and intuitions; and then the further distinction between pure, and empirical intuition. But the move to the representation of this a priori particular as infinite, such that it guarantees particular constructions in space in its turn, requires the extended idea of intuition as related to the apparatus of sensibility. Space becomes the formal principle of the sensible world, which contains in it all objects that can come before us as objects of experience.

Nonetheless, it is hard to be quite clear about this particular Kantian claim. What Kant says about spaces and times always being given as limited, but at the same time limited by more general spaces and times, suggests the conclusion that space and time are unlimited given magnitudes, rather than infinite. But at this point in the argument Kant's empirical realism is complemented—some might say denied—by his transcendental idealism. What is "given" is not a simple experiential fact, but an integral part of transcendental philosophy and the a priori conditions which make this possible. Thus when Kant asserts that space and time are a priori intuitions, this must be held to imply the fact that what can be said of our experience is proscribed by these very conditions, and not so much that they describe it. To say that space and time are infinite given magnitudes, therefore, does not mean that they are given as infinite, but that the very possibility of their being *given* rests on certain conditions, one of which is that the transcendental concept of infinity is presupposed.

The residual effect exercised on Kant's thinking by the demands and presuppositions of Euclidean geometry on the one hand, and Newtonian absolute space on the other, is clear in the discussion of space and time in the Aesthetic. Although Kant is concerned to offer a theory of space in opposition to Newton (and Leibniz), the presuppositions of a space which is in effect one in which Euclidean geometry applies and is true, remains always just below the surface in this important part of the argument. Kant is prepared, like Newton, to deny that we can imagine a

CHAPTER TWO – KANT'S THEORY OF SPACE AND TIME

space which has all parts of space as its parts. (Newton believed that such a space exists, and that we can understand—but not imagine—that such a space exists.)

It seems, then, that the assumption of the infinitude of space is a necessary element in the analysis. This can be expressed in that somewhat oblique way I gave earlier; that it is necessary to *postulate* infinite space, but space is not necessarily infinite. Though our experience is necessarily only of finite spaces, a theory of space as pure intuition and a priori representation could not be carried through on such a basis, even if the "indefinite" magnitude of space was admitted as the only theory which kept within the limits of the experiential data. Here, as throughout the critical philosophy, Kant is trying to demonstrate the inadequacy of giving a full description of experience without the assumption of non-empirical principles which make this experience possible.

The metaphysical exposition of time is analogous to that for space. Our representation of time could not be acquired empirically, since neither co-existence not succession, with which we are acquainted through experience, are possible without the presupposition of time itself. Time is a necessary representation that underlies all our intuitions, both outer and inner, we cannot "think away" time, *(aufheben)*, though we can think of it as empty of outer appearances. Kant concludes from this that time must be given a priori.

It is clear enough that this argument parallels that for space, but the strict parallelism runs in to difficulties here. Time is the "immediate" form of inner sense, and the "mediate" form of outer sense; that is, although not all objects are in space (e.g. thoughts), all objects are in time. Thus we can think of a space empty of objects, thereby separating in our mind the form of intuition from its content. But how can we "think away" time without thereby removing the thought itself which, as the "object" of our inner sense, is conditioned by time? So to think an empty time is impossible; thoughts are necessarily in time and thereby given time a content. The only plausible interpretation of this passage seems to be that appearances, as spatio-temporal or temporal, may all "fall away", (*wegfallen*); but time cannot be removed from appearances without them ceasing to be appearances. What Kant says in this context in the Aesthetic should be compared to what he says much later in the First Critique, (in B225), where he writes that all appearances are in time, which is the condition for their perception as successive or co-existent; yet time without this content is "a merely ideal thing".

The problem for us here is that the Aesthetic is not completely comprehensible without the teachings of the Analytic. Kant has made it clear in the first part of the Aesthetic that he is "isolating" sensibility in order to leave nothing for analysis but intuition. This process of isolation is quite artificial and leaves us the forms of intuition, which are "pure space and pure time, (*ens imaginarium*)". [A291]

Kant tells us that space is not a real object which can be outwardly intuited: "Empirical intuition is not a composite of appearances and space (of perception and empty intuition). The one is not a correlate of the other in a synthesis; they are connected in one and the same empirical intuition as matter and form of the intuition". [B457]. He says in the same passage that we cannot set space outside all appearances without giving rise to "all sorts of empty determinations of outer intuition", which are not possible perceptions. Does this not flatly contradict the Aesthetic, where Kant has told us that we can indeed think space as empty of objects? The contradiction is apparent, not real, because for the purposes of the Aesthetic, that is, for the purposes of finding out what belongs to a concept, we can treat the form of sensibility as if it was separate from the matter of sensation. But we really know only spatio-temporal appearances, and these only can be objects of a possible experience. Our mode of intuition is dependent on the existence of the object and is thus possible only if the subject's faculty of representation is affected by that object. If we take space and time, the formal conditions of experience (*ens imaginarium*), as objects, Kant says we will become involved in "empty determinations' which transcend all possible experience. This was Newton's error. We can treat pure space and pure time as if they were real, but only for the purposes of analysis of what belongs to a concept.

Newton, perhaps for "theological" reasons, could not resist treating these "Ideas of Reason" as real, self-subsisting entities. Thus Kant retains the regulative function of absolute space, without thereby committing himself to hypostatising it as a metaphysical reality.[9]

Kant claims to have shown that time is an a priori intuition, where the parts of time are only regarded as enclosed between limits in an all-embracing time series. From this, of course, follows its infinitude. This line of reasoning closely parallels the argument for the infinitude of space, although we can find, in the *Inaugural Dissertation*, a remark clarifying the example of time. Kant says there that time is a continuous quantity—any part of time is itself a time; and instants are not parts but limits. "For when two moments are given, time is not given except in so far as in those moments actual things succeed one another. Therefore in addition to a given moment there must be given a time in whose later part there may be another moment". [P.C. p.64]

Perhaps the most that Kant can claim to have shown is that space and time are given as unbounded; that is, beyond any determinate space whatsoever we must always postulate a further space. But unboundedness is not infinitude—physics and mathematics gives us spaces that are unbounded but finite. Kant could have made a claim such as this consistently with his general doctrine of space and time, but his acceptance of Euclidean geometry commits him to the much stronger idea that space and time are "given" as infinite, notwithstanding the constraints upon how this is to be understood as outlined above. It has been suggested (for instance by Walsh) that what Kant meant by "infinite given magnitude" was really that space and time are infinitely divisible. No doubt this is Kant's view, but this is only a corollary to his views on space and time qua infinite in the sense of containing determinate spaces and times. If one accepted Walsh's reading, Kant would be committed to space and time as a priori intuitions (particulars), unbounded in the sense

---

[9] This is one way in which Kant frequently gets discussion of old problems moving again, viz. by "...converting metaphysical principles into something possessing purely methodological force". See Buchdahl, (1969), p.511.

that they are conditions for the possibility of determinate spaces and times, and hence are larger than any such magnitudes; they are also continuous and infinitely divisible. We might want to say that the idea of an infinitely divisible space and time is as problematic for the Aesthetic as infinitely large spaces and times: both claims follow from their nature as transcendental conditions for appearances—neither are, in any obvious sense, experienced as such. They are "tasks" that reason sets itself; they are *"Grenzbegriffe"*, Ideas of Reason. Space and time are absolute for Kant in this sense only, rather than in Newton's, who sees them as infinite, self-subsisting entities. The latter idea was regarded by Kant with scorn, who refers to them as "monstrosities", (*Undinge*). For Kant they are really only "schemata for coordinating with each other absolutely all things externally (and internally) sensed".

The metaphysical expositions have tried to show that the empirical reality of space is grounded not in the nature of things-in-themselves, nor in the existence of any absolute, self-subsisting entities, but in the focal conception of possible experience. The pervasive and universal feature of appearances, viz. their spatio-temporal nature, is guaranteed by the formal conditions which enable us to have these experiences in the way we do in fact have them. Space and time are forms of intuition, and yield spatiality and temporality as such. The geometrical (and "chronometrical") structure of this bare spatiality and temporality is, however, a matter for the transcendental expositions. The synthetic a priori nature of the principles of geometry can, according to Kant, be "deduced" from the nature of space and time as a priori particulars.

I noted above that the interdependence of Kant's theory of space as a priori intuition, with his theory of geometry, is sometimes taken as being so close as to entail the rejection of one should the other be considered as inadequate. This is not obvious, but no doubt the development of alternative geometries must influence our assessment of the Aesthetic to some extent. This takes us to the very heart of Kant's philosophy of mathematics, and is the subject of the following section.

2.5 Construction and Schematism

The idea that Kantianism in the philosophy of mathematics has ceased to be tenable because of the very existence of non-Euclidean geometries is widely held. It is true that the development of hyperbolic and elliptic geometries compels a re-assessment of some of Kant's claims about the supposed science of "real" space. Yet the rigidity of Kantianism in this area can be exaggerated, and the overall subtlety of Kant's position too easily overlooked. Kant's position not only explicitly allows alternative logical possibilities, but implicitly demands their existence. [See Brittan, p.68]

The distinction Kant draws is between that which is merely logically possible, and that which we can call "constructibility", where the latter term is to be understood in relation to pure intuition. Kant is often interpreted as saying that constructions in the a priori intuition of space are indispensable for geometrical science. It would follow from the acceptance of this view that since both analytic and non-Euclidean geometries dispense with spatial figures (though they may be used heuristically, or "analogically"), then Kant's theory of geometry, if not his entire philosophy of mathematics, must be abandoned as anachronistic.

The emphasis that Kant places on spatial figures must seem almost naive—especially since the Cartesian programme of analytic geometry can hardly have escaped his attention. Indeed, this presents us with an interesting problem: how does the existence of algebraic geometry affect the orthodox (and obsolete) view that Kant is presumed to hold, viz. that spatial figures are indispensable? Descartes' fundamental idea was to employ geometric intuition to elucidate algebraic relationships. But how, precisely, are we to understand what Kant took to be the epistemological connection between analytic and "synthetic" geometry? One answer might be that he could argue that the qualitative essence of figures cannot be reduced to numerical—hence algebraic—representations. He could argue, for instance, that analytic geometry is an analytic representation of the quantitative determinations of figures, yet still do not capture the essence of such figures as spatial. This response would presumably chime with those who argue that Kant's theory insists that Euclidean geometry is a description of our spatial intuition.

Another way of grasping this relationship between Kant's theory, and analytic geometry, is provided by the Transcendental Doctrine of Method, where Kant makes some of his most suggestive and important comments on the idea of construction. It is there that Kant distinguishes philosophical from mathematical reasoning: the former proceeds by "reasoning from concepts"; the latter is "synthetic", finding its classic exemplification in the axiomatic method of Euclid, where theorems are deduced from axioms by strictly logical reasoning. Thus philosophy is reasoning from concepts: mathematics is reasoning from the construction of concepts. Algebraic geometry, on the other hand, proceeds analytically. Since Kant was convinced that his discovery of this distinction in the respective methods of philosophy and mathematics was of major significance for our understanding of both, the Euclidean synthetic method must have seemed a crucial example of this fact. This would account for the emphasis he gives to the synthetic geometrical method, and to the concomitant neglect of the analytic method for his purposes; that is, for the purpose of founding metaphysics as a science which would offer the same certainty as geometry.

A third possibility, which is in some respects the most interesting, is that Kant in fact regarded algebra as in some sense more fundamental than either arithmetic or geometry, and that Cartesian geometry simply realises this idea in a way that Kant's own theory can accommodate. But does Kant have a theory of algebra that could support such a reading? There is no easy answer to such a question, given that an emphasis on spatial figures pervades much of Kant's discussion on this subject. Of course one could simply argue that since the context of much of this discussion is the Transcendental Aesthetic—explicitly concerned with space and time—it would be inappropriate for Kant to consider the reduction of spatial to algebraic relationships. In Kant's theory, after all, spatial figures still apply to the space of perception, and it is this descriptive quality of geometry with which Kant is concerned in the Aesthetic.

It is generally accepted that Kant's theory allows at least the logical possibility of alternative geometries. The idea of construction is a constraint on "real" geometries—interpreted systems that purport to apply to the space of experience. I shall argue that by emphasising what we understand by construction within the critical philosophy, then linking this to the difficult doctrine of the schematism, we can present a plausible theory of algebra which Kant could have accepted.

There are primarily two considerations—one direct, one indirect—that support the thesis that Kant's philosophy of mathematics allows for the existence of alternative geometries. The indirect consideration is that since Kant insists on the synthetic character of geometrical propositions, the replacement of the parallel axiom by its contrary would generate no inconsistencies in the system taken as a whole. Since this is obviously the case with non-Euclidean systems, it would follow that Kant is correct in asserting the non-analyticity of the axioms and postulates of Euclidean geometry. We may leave this indirect consideration without further comment, since nothing in what follows depends upon its acceptance as valid.

The direct consideration rests principally on this passage from the First Critique:

...whence shall we derive the character of the possibility of an object which is thought through a synthetic a priori concept, if not from the synthesis which constitutes the form of the empirical knowledge of objects? It is indeed a necessary logical condition that a concept of the possible must not contain any contradiction; but this is not by any means sufficient to determine the objective reality of the concept, that is, the possibility of such an object as is thought through the concept. Thus there is no contradiction in the concept of a figure which is enclosed within two straight lines, since the concepts of two straight lines and of their coming together contain no negation of a figure. The impossibility arises not from the concept in itself, but in connection with its construction in space, that is, from the conditions of space and its determination. [B268]

Here, Kant is identifying mathematical existence with the possibility of construction. Any mathematical object—in this case a geometrical figure—"exists" in so far as it can be constructed in pure intuition. Generally, Kant is taken to mean that space—experiential space—is given to us as definitely and irrevocably Euclidean. This amounts to the assertion that perceptual space, the space of any and every possible experience, could not be "re-constructed" such that non-Euclidean geometry might provide the formal basis for intuitive, i.e., particular, constructions in space from which synthetic judgements valid a priori could follow.

In addition to being absolutely central for an understanding of Kant's views on the nature of mathematics, this idea of construction occupies a pivotal position from which to integrate Kant's theoretical and aesthetic philosophy respectively.[10] So what can we say is involved in this idea? We start by emphasising that the synthetic propositions of geometry are "objectified", so to speak, and thus capable of being verified, by constructing the object of the concept in pure intuition; that is, by "exhibiting" a priori the intuition which corresponds to the concept. (This idea of a correspondence between the intuited object, and the concept is, I believe, crucial: I will say more of this below.) For Kant, the test of a "real" geometry is this appeal to the possibility of constructing its figures—its objects—in pure intuition. What this amounts to is the "exhibition" of particulars which manifest features true of a whole class of entities. We can say that for Kant, pure constructions in space (and time) are *symbolic instantiations*. It has been cogently and consistently argued by Jaako Hintikka that a proper understanding of Kant's philosophy of mathematics depends upon recognising that there are two

---

[10] An imaginative and provocative discussion of the relationship between Kant's philosophy of mathematics and his aesthetic theory is provided by Donald Crawford, in his "Kant's Theory of Creative Imagination", in Cohen and Guyer, p.151 ff. I also extend this discussion in my "Art and Mathematics in Kant's Critical Philosophy", (1988).

distinct but related notions of "intuition" in Kant. What Hintikka calls the "mature" theory links intuition with sensibility directly. This meaning is, however, easily confused with a more restricted and original meaning, to be found in the pre-critical writings, and also in the Discipline of Pure Reason. In this case, "intuitive" means that which represents an individual, to be contrasted with the notion of a general concept. It is not the spatial character of intuitive constructions that is of crucial importance, but more the fact that they can be employed as exemplars for a general class. The constructed figure, for instance a triangle, is the spatial representation of the abstract relations which constitute triangularity. The point is then that such a figure is useful to us precisely because it embodies those relations which are less easily grasped independently of it.

The a priori exhibition of a concept by means of an intuitive construction may amount to nothing more complicated than an empirical procedure such as making marks on paper, or manipulating the beads of an abacus. A more sophisticated but quite plausible interpretation of what Kant is getting at here with this idea of construction qua exhibiting in intuition, is by means of the logical procedure of existential instantiation. In this sense, a construction is a particular which might be thought of as a concept "made flesh". Construction is a general way of allowing the deduction of F(a) from the existentially quantified proposition $(\exists x)(Fx)$.

The test of the meaningfulness of a concept, of its real rather than merely logical possibility, is the construction of a figure, produced a priori, in a way that bears a formal similarity to syllogistic reasoning—in other words, the determination of particular conclusions under general rules, by means of the faculty of judgement. Yet it is at the same time an appearance present to the senses. [A240/B299] This idea—that Kant's philosophy of mathematics can be "reconstructed" by appeal to quantification theory—is at the heart of Hintikka's interpretation. The use of the natural deduction rule of existential instantiation introduces new representatives of individuals: this, on Hintikka's view, is precisely what Kant's use of intuitive construction involves, and it pre-dates the meaning of intuition given in the Aesthetic, where the connection with spatial intuition is made directly. For Hintikka, the idea that the mathematical method is based on the use of general concepts *in concreto*—that is, in the form of individual instances—provides the starting-point for Kant's mature theory of mathematical reasoning. [Hintikka, (1965); (1967); (1969).]

In a very general way Kant's view may be identified with his claim against rationalist metaphysics that "existence" is not a predicate:

...all existential propositions are synthetic...Anything we please can be made to serve as a logical predicate; the subject can even be predicated of itself; for logic abstracts from all content. But a determining predicate is a predicate which is added to the concept of the subject and enlarges it.[A598/B626]

We will see later that for Kant the function of schemata is to "particularise" certain concepts, or to present in intuition individuals that represent a general class.

It would be wrong to assume that Kant thinks that this figurative construction is somehow complete in itself. The process of construction must not be thought reducible to an empirical procedure, valid for the spatial figure presented but only for this figure. This would fail to yield the characteristics bound up with the recognition of mathematical truth, viz. necessity and universality. Kant often seems to assume that any theory of mathematical judgement which failed to account for, or incorporate, this necessity and universality could not, for that very reason, be accepted. The argument seems to be that given these characteristics, what theory would account for the judgements of mathematics being also non-analytic? The non-analyticity of such judgements is not given as a condition of their universality and necessity, but as part of a wider theory that assumes these very conditions. To obtain synthetic propositions about triangles it is not, in Kant's view, sufficient merely to consider the concept "triangle"; this would yield only analytic propositions. However, if we exhibit the triangle in intuition—if we draw a triangle, or think of one in imagination—then such a construction putatively generates that body of synthetic propositions, valid a priori, with which Euclidean geometry has made us familiar.

Kant seems to be assuming here that imagination has some irreducible connection with the production of mental images; yet I will argue later that it is one of Kant's achievements to have made imagination a more formal kind of faculty than this suggests. But it is no doubt correct for Kant to affirm that the production of mental images is one of the functions, at least, of imagination, if not perhaps the philosophically most interesting one. What this does show is that there must be something more to the notion of construction than the production of lines on paper or images in the mind's eye. And indeed Kant does go on to provide that additional feature.

In order that the constructed figure is what Kant calls "adequate to the concept", the procedure is explained in terms of transcendental imagination, that is, in terms of a priori conditions. Kant must build into his account some element that is presuppositional: in this way the empirical construction is given an a priori component. With this in mind, we can return to our earlier question: how can we be certain that what is "read" from the individual figure is valid for all possible figures of this kind? Kant's answer is that by employing the imagination to construct a triangle in pure intuition we uncover through "regressive" analysis the a priori conditions by which imagination is itself bound in producing figures of this kind.

> The single figure which we draw is empirical, and yet it serves to *express* the concept, without impairing its universality. For in this empirical intuition we consider only the *act* whereby we construct the concept, and abstract from the many determinations...which are quite indifferent, as not altering the concept 'triangle'." [A714/B742, my emphasis.]

This consideration of an "act", which is presupposed in the empirical construction, is what supplies the essential presuppositional element. Mathematics, Kant insists, does not extend knowledge by the analysis of concepts alone: verification in mathematics requires that we "hasten to intuition". In pure intuition the concept is instantiated and considered *in concreto*, yet non-empirically, since the construction is in pure, not empirical intuition. The concept is "particularised", constructed; and whatever follows from the universal conditions of the construction is universally valid of the object of the concept thus constructed.

In order to produce a particular construction adequate to the concept, we require some form of mediation between understanding—which for Kant is the faculty of rules which at the same time provides a priori concepts—and sensibility, in whose domain the constructions must be presented if they are to acquire existential significance, that is, sense. In this way, we effect an isomorphism between the a priori truths belonging to the concept "triangle", and the identifiable a priori conditions exemplified in the construction. It should be remembered that in Kant's hierarchy of faculties, it is *judgement* that has the task of subsuming under rules; in general, it is that procedure that moves us from a major to a minor premise of a syllogism to a particular conclusion. Thus Kant introduces, as part of the Transcendental Doctrine of Judgement, his idea of the Schematism of the pure concepts of understanding: "If understanding in general is to be viewed as the faculty of rules, judgement will be the faculty of subsuming under rules". [A132/B171]

It is the notoriously difficult Schematism chapter which expands the implications of mathematical construction, and contributes to a less constricted understanding of Kant's philosophy of mathematics.

We have seen that the figure produced in intuition from which synthetic propositions, valid a priori, may be "read off", must in some way be representative of all figures of that kind. Any characteristic possessed uniquely by the empirical figure can be abstracted and ignored in the reasoning process. But how can a single figure perform such a task adequately? As Kant readily concedes, no *image* could be adequate to the general concept "triangle".[11] Kant's answer is to be found in the doctrine of the transcendental schematism.

At this point it will be helpful to clarify the role of "synthesis" within the critical philosophy. This will prepare us for the special use Kant makes of this idea in the schematism itself. We must be clear that this idea of construction is much wider than our focus on the philosophy of mathematics might indicate. Certainly it is located, initially, inside such a framework, but it can be generalised as the process of synthesis of the empirical manifold. [See Buchdahl, (1969), p.556.] As I have suggested, this synthesis provides the presuppositional or transcendental element through imagination. The connection between concepts and intuitions is effected by means of a synthesis of which the schematism is the focal example:

> Synthesis in general...is the mere result of the power of imagination, a blind but indispensable function of the soul, without which we should have no knowledge whatsoever, but of which we are scarcely ever conscious. To bring this synthesis *to concepts* is a function which belongs to the understanding, and it is through this function of the understanding that we first obtain knowledge properly so-called. [A78/B103][12]

This is the mirror-image of the specific case of mathematical construction. In mathematics we produce an image for a concept, by means of what Kant calls an "imaginative synthesis":

> The image is a product of the empirical faculty of reproductive imagination: the schema of sensible concepts, such as figures in space, is a product and, as it

---

[11] This clearly reminds us of Berkeley's struggle against "abstract general ideas". Cf. also Buchdahl, (1969), p.285: "When a geometer appears to 'reason round a triangle on a blackboard', he is not drawing general conclusions from a particular triangle; but neither is he 'really thinking' of some 'abstract universal triangle'. Rather, he is reasoning about those properties of his triangle which are held in common with the class of triangles concerned in the demonstration, that is to say, with all those triangles which are given the properties mentioned in the definitions, postulates, and axioms".

[12] Kemp Smith's emphasis. The language here reminds us of the Schematism itself. Cf. A141/B181.

were, a monogram, of pure a priori imagination, through which, and in accordance with which, images themselves first become possible. [A141/B181]

To subsume particulars under concepts is the task of the faculty of judgement in general, and the schematism in particular. The productive synthesis of imagination is a transcendental act:

> We cannot think of a line without *drawing* it in thought, or a circle without *describing* it. We cannot represent the three dimensions of space save by *setting* three lines at right angles to one another from the same point. Even time itself we cannot represent save in so far as we attend, in the *drawing* of a straight line (which has to serve as the outer figurative representation of time), merely to the act of the synthesis of the manifold whereby we successively determine inner sense, and in so doing attend to the succession of this determination in inner sense. [B154]

Thus connection, or synthesis of the manifold, is not some merely passive process undertaken through sensibility and intuition: it is an active procedure of the faculty of imagination. Time—as *formal* intuition, demands synthesis of imagination qua transcendental act. As *form of* intuition, time is the undifferentiated phenomenon of lapse, yielding only the possibility of determinate succession. [See Buchdahl, (1969), p.642.] The successive synthesis of the manifold—an act performed by means of the productive imagination—locates this entire problem for Kant within transcendental philosophy. [B155 & note; A163/B204] Geometry itself, as the "mathematics of space" (*Ausdehnung*) is grounded in the productive imagination in the generation of figures. It is on this basis that axioms are understood as conditions of a priori intuition in figurative construction.

In the transcendental deduction Kant had argued that the pure concepts of understanding apply to the objects of intuition in general. For this reason however, these concepts are incapable of giving determinate knowledge of objects. It is schemata that have the task of "particularising" concepts in the required sense. Only the schematism, qua transcendental act, can provide this determinate knowledge of objects. The schema is a product of imagination: it is a universal procedure—an *act*—which provides an image for a concept. In Kant's words, it is "...a rule of synthesis of the imagination, in respect to pure figures in space". [A141/B180] The schema of a sensibilised concept, in this case a figure in space, is a product of pure a priori imagination through which, and in accordance with which images—something empirical—first become possible. Speaking "transcendentally" therefore, it is not the constructed triangle as such which is the ground of a priori valid synthetic propositions, but rather the fact that it was produced in accordance with the schema for "triangle", either as a figure on paper or imaginatively.

Images are connected with the concept by means of the schema which they designate. This schema for "triangle" is a rule of procedure for construction in intuition. Without such an a priori rule of construction we could not be certain that we had in fact produced a triangle. The drawing is a particular which presents an instance of the class "triangle", and thus *represents* this class; this enables us to recognise the figure qua instance of a geometrical class rather than, say, an undifferentiated spatial area, or any of the various other possibilities implicit in the empirical construction.

This suggests that the possession and application of a schema, though essential for the recognition and construction of figures such as triangles, cubes, etc., would not be necessary for the recognition and "classification" of more "primitive" topological figures, where the properties of being a closed curve, for instance, are more important than the kinds of conceptual characteristics of Euclidean figures that Kant has uppermost in his mind. As he points out, in mathematics we consider the universal *in* the particular,

> ...or even in the single instance, though still always a priori and by means of reason. Accordingly, just as this single object is determined by certain universal conditions of construction, so the object of the concept, to which the single object corresponds merely as its schema, must likewise be thought as universally determined. [A714/B742]

An immediate objection to this idea might be that it is in an important sense superfluous. Kant needs an "image", a spatial figure produced in intuition, in order that a priori valid, synthetic propositions, may be read off from it. But could it not be argued that the empirical construction serves merely as an heuristic aid, and is not a necessary element of the reasoning process? The schema, as a rule of procedure for constructing any image for a concept, must "contain", in the abstract, or "pre-constructively", so to speak, all of the information that can in principle be included in, and thus read off from, the intuitive construction qua particular instantiation. If this were not the case, the constructed figure could not be "adequate to the concept"; that is, there would be either more, or less "information" in the empirical figure than is in the concept. This rule of construction should therefore contain, in principle, all that the geometer requires in order to reason about triangles etc. More precisely, it would suggest a geometry which dispenses altogether with spatial constructions.

The idea of such an act of imagination can be understood only within the context of the notion of synthesis alluded to above. Nonetheless, even for Kant it seems miraculous how such functions of the imagination can be the foundation for a system of relations which, when spatially interpreted, generates an a priori science which has application to experience; whilst yet nothing much can be said of it except that it exists. In fact, in talking of the schematism, Kant is at one point reduced to conceding that it is "...an art concealed in the depths of the human soul, whose real modes of activity nature is hardly likely ever to allow us to discover, and to have open to our gaze". [A141/B181; B103]

Nevertheless, this conception of the schematism implies that geometrical science could dispense with spatial constructions. Significantly, however, it could not dispense with *temporal* constructions. Time—as the form of inner sense—is the necessary condition of all experience; that is, outer, or spatial experience; and inner experience—minimally temporal and maximally spatio-temporal. Synthesis of the manifold of pure a priori intuition gives knowledge of objects: this

CHAPTER TWO – KANT'S THEORY OF SPACE AND TIME

synthesis, this "taking up and connecting", is the result of the transcendental procedures of imagination. As a function effecting subsumption of intuitions under concepts, it is a task for transcendental schematism. "If this manifold is to be known, the spontaneity of our thought requires it to be gone through in a certain way, taken up and connected. This act I name synthesis". [A77/B102]

This is best understood in relation to Kant's definition of number. To think a number "in general" is the representation of a method whereby a multiplicity may be represented in an image in conformity with a concept. In Kant's cryptic formulation, number is "...simply the unity of the synthesis of the manifold of a homogeneous intuition in general". [A143] The movement of consciousness produces undifferentiated succession in the manifold of inner sense: synthesising the manifold is "taking up and connecting". And number in general is the presented product of such a synthesis. [A103] So Kant thinks that a number is simply a conventional way of marking a determinate position in the manifold of inner sense. He had maintained this position as early as the pre-critical writings, where he suggested that numbers are to be regarded as "sensuous epistemological tools". [P.C. p.24]

It is important that we keep in mind that schemata are not themselves spatial images—much confusion is generated by thinking of them as such. They are a priori determinations of time in accordance with rules which make images possible. [A145/B185] In this way, we can locate a Kantian "pure science of time" within transcendental philosophy. The point is that time is more general—less dispensable—than space: the "science" of time must therefore be more fundamental than geometry qua science of space. The pure science of time cannot be arithmetic, since arithmetic has actual numbers as its objects and is insufficiently general. What we are looking for is a science of number in general which, through its connection with the fundamental transcendental synthesis of the manifold of inner sense, concerns "taking up and connecting" in an arbitrary way. Such a science would be the condition for the possibility of both arithmetic and geometry: this science is algebra.

Mathematics does not only construct magnitudes (*quanta*) as in geometry; it also constructs magnitudes as such (*quantitas*), as in algebra. In this it abstracts completely from the properties of the object that is to be thought in terms of such a concept of magnitude...Once it has adopted a notation for the general concept of magnitudes so far as their different relations are concerned, it exhibits in intuition, in accordance with certain universal rules, all the various operations through which the magnitudes are produced and modified. Thus in algebra, by means of a symbolic construction, just as in geometry by means of an ostensive construction, we succeed in arriving at results which discursive knowledge could never have reached by means of mere concepts. [A717/B745][13]

In his theory of geometry Kant appears to insist on the indispensability of figures in space. Yet the teaching of the schematism, emphasising as it does the fundamentally temporal nature of rules of synthesis for generating figures in space, links algebra to the intrinsically temporal character of construction by means of the symbol. Kant's theory suggests, albeit indirectly, that spatial constructions are dispensable, provided we are in possession of an adequate system of symbols by means of which any intuitive—that is, particular—relations, can be expressed. The algebraic method is not geometrical, but it is constructive in the required sense: that is, it employs variables, the only acceptable value of which are individuals. [Cf. Hintikka, (1967), p.359: and A159/B198]

The concepts expressed through and instantiated in the symbols, especially those concerning relations of magnitude, are presented in intuition: they are *symbolically instantiated*. The *sine qua non* of geometrical science is not the existence of spatial figures, but the construction in pure intuition, i.e. the possibility of considering the universal in the particular construction. [Cf. Brittan, p.53 ff.] This may be either a spatially extended figure, or an algebraic representation of the relations expressed in such a

---

[13] Although a detailed discussion of it would take me outside the scope of this work, we should note the interest in W.R.Hamilton's work on algebra as the "science of pure time", and its relationship to Kant. In particular, see T.L.Hankins; J. Hendry; Peter Øhrstrøm; and my (1982).

figure. Kant refers to algebra—"universal arithmetic"—as an ampliative science, insisting that the remaining parts of pure mathematics (*mathesis*), progress largely because of algebra considered as a universal theory of quantities. As Hintikka has pointed out, Kant's theory of mathematical reasoning, and especially the interpretation of intuition which emphasises its non-spatial character, can be identified in the pre-critical phase. As early as 1763, Kant had distinguished mathematical from metaphysical reasoning by means of the former's use of signs, known "individually and sensibly", which give concrete knowledge of general concepts. [P.C. p.13 and p.24.]

This interpretation of construction and schematism is consistent with what explicit remarks on algebra there are to be found in the *Critique of Pure Reason*. That said, there does seem to be a serious exegetical problem raised by some typically convoluted comments made by Kant in the *Critique of Judgement* which bear directly on this issue and, indeed, suggest a fundamental inconsistency in Kant's use of key terms.

In section 59 of the Third Critique, Kant draws some distinctions between *schemata* and *symbols* which cannot easily be reconciled with his more detailed remarks on the use of notation made in other places. In that section, Kant says that all concepts demand verification by means of intuitions: this is part, at least, of what is meant by Kant's affirmation that thoughts without content are empty, and intuitions without concepts are blind. Neither concepts in the absence of a corresponding intuition, nor intuition without concepts, can yield knowledge. Empirical concepts are verified by examples; pure concepts by schemata. This process of verification, or "rendering in terms of sense" as Kant puts it, can occur in one of two modes:

> Either it is schematic, as where the intuition corresponding to a concept comprehended by the understanding is given a priori, or else it is symbolic, as where the concept is one which only reason can think, and to which no sensible intuition can be adequate. In the latter case the concept is supplied with an intuition such that the procedure of judgement in dealing with it is merely

93

analogous to that which it observes in schematism. In other words, what agrees with the concept is merely the rule of this procedure, not the intuition itself. [C.J. p.221]

So far, so good. Where the concept is an Idea of Reason such that there is in principle no intuition that could be adequate to it, the expression of the concept is made by means of a symbol. (An "Idea of Reason" we understand as a concept neither abstracted from nor applicable to, sense-experience: it "transcends the possibility of experience".) The relationship between a symbol and its concept is merely analogous to the manner in which a schema relates to its concept. Both the schematic and the symbolic are, for Kant, intuitive modes of representation. The difference is that the former directly present the concept through demonstration; the latter, on the other hand, are merely indirect presentations of the concept by means of analogy.

Such an interpretation of symbolism is what one would expect, given the critical philosophy's insistence on the transcendent character of certain concepts of reason. It is clear that such concepts could only be given intuitive—and hence, immanent—meaning, through analogies of some kind. However, Kant then identifies both schematism and symbolism as "hypotyposes", that is, as "presentations", (*Darstellungen, exhibitiones*), and not mere marks, (*Charakterismen*). Marks are

> ...merely designations of concepts by the aid of accompanying sensible signs devoid of any intrinsic connection with the intuition of the object. Their sole function is to afford a means of reinvoking the concepts according to the imagination's law of association—a purely subjective role. Such marks are either words or visible (algebraic or even mimetic) signs, simply as expressions for concepts. [C.J. p.221][14]

---

[14] In the Second Critique, Kant warns against the transformation of a symbol into a schema: this, he says, "leads to mysticism" in practical reason. See *Critique of Practical Reason*, p.162.

This seems to present a serious problem. Here, Kant is identifying algebraic symbols as merely conventional marks, the purpose of which is to reinvoke concepts by means of simple association. Earlier, I suggested that algebraic expressions directly presented in intuition relations of magnitude as such, so that they could be connected to the rules of synthesis described as schematism. Yet here we have Kant apparently locating algebraic notation inside a wider concept of *symbolism*, rather than insider a wider concept of schematism. The relationship between an algebraic symbol and a number concept should be direct, and is quite different from the relationship that a model or analogy has to that concept of reason for which it is such. The connection between a symbol qua analogy and its concept is looser than the connection between schemata and concepts. As Kant points out, one thing may be used as a symbol for another by virtue of the similarity in the "structure of reflection" in the two case:

> In this way a monarchical state is represented as a living body when it is governed by constitutional laws, but as a mere machine (like a hand-mill) when it is governed by an individual absolute will; but in both case the representation is merely *symbolic*. For there is certainly no likeness between a despotic state and a hand-mill, whereas there surely is between the rules of reflection upon both and their causality. In language we have many such indirect presentations modelled upon an analogy enabling the expression in question to contain, not the proper schema for the concept, but merely a symbol for reflection. [C.J. p.223]

Thus symbols qua analogies may express concepts for which the direct employment of an intuition is out of the question. In the Third Critique the idea of representation by means of analogy is used as Kant's designation of symbolism; although this idea is clear enough, algebraic symbolism should not be found within its compass.

An explanation for this confusion may be found in the Third Critique's concern for so-called "reflective" judgement, to be contrasted with determinant judgements. Kant says that if a universal in the form of a rule, principle, or law is given, then the judgement which subsumes the particular instance under it is a determinant judgement: if, on the other hand, the particular instance only is given, reflective judgement concerns finding a universal for it. Determinant judgement, i.e. the subsumption of particulars under rules operates, according to Kant, "...even where such a judgement is transcendental and, as such, provides the conditions a priori in conformity with which alone subsumption under that universal can be affected". [C.J. p.18] So unlike transcendent judgements, where ideas of reason can be represented only through analogies; transcendental judgements, involving as they do the a priori conditions of knowledge, can be made determinate by means of schemata. Now since rules of synthesis of a priori imagination are presupposed in all construction of mathematical objects, such objects must present their concepts, and make them determinate, in a way that is quite different from reflective judgement by means of symbols which, as Kant says, represent by "mere analogy". We conclude therefore that Kant's general theory of mathematical construction militates against any consideration of algebraic notation as mere marks, notwithstanding that such notation is "conventional". These symbols are important for Kant as practical devices, but the a priori locus of mathematical construction is the procedure of imaginative synthesis. [Cf. Heyting, p.70]

Kant's idea of schematism as the procedure through which a construction in intuition becomes an "embodied" concept is the focal point of what I have called his "anti-logicist" programme. The implications of this analysis can also help us to grasp the full implications of some further—and almost equally controversial—aspects of his philosophy of mathematics. However, it will be necessary to postpone discussion of how the issues raised until this point contribute to our understanding of these other matters, until they have been separately analysed in what follows.

CHAPTER TWO – KANT'S THEORY OF SPACE AND TIME

2.6 Spaces and Geometries

So far as Kant is concerned, "space" means "Euclidean space" only in the sense that the determinate space of physics and mathematics is the result of the constructive procedures proscribed by the forms of intuition, and that these constructions will be found to accord with the axioms and postulates of Euclidean geometry. Kant never considered seriously the possibility that perceptual space could, for us, be correctly described by any alternative geometry. We might resile from Kant's view by in fact denying that the idealisations of *any* geometry are ever instantiated in perception. For example, there simply are no Euclidean triangles available for empirical inspection. Their "being" consists in their ideal meaning within a formal system. The propositions of pure geometry are neither confirmed nor refuted by experience. It may be necessary that, in order to reason by means of geometry, we must impose some particular geometry or other on the undifferentiated space of perception, but this is a much weaker claim than Kant seems often to want to make.

The fruitfulness of non-Euclidean geometry is unquestioned, forming as it does part of the structure of a whole new system of physics. But how is this fact to be explained? The question of how an abstract system can have physical application is as difficult to answer in this case as in the application of Euclidean geometry to Newtonian physics. By their nature, pure geometries qua uninterpreted formal systems, do not apply at all to empirical objects, though in certain circumstances it can be useful to identify the latter with ideal structures of one kind or another.[15] Whatever meaning we may give to non-Euclidean geometry for modern physics, for the world of empirical objects it is often regarded as meaningless to claim that this space and its relations is either Euclidean or non-Euclidean, although it may sometimes be convenient to treat Euclidean geometry as if it was true of terrestrial space and its objects; and conversely, it may be

---

[15] The notion of an "idealising identification" of empirical and mathematical objects is developed by Körner (1966), *passim*.

convenient to treat some non-Euclidean system as if it was true of certain more remote objects of cosmology.

Kant is criticised for his views on geometry in two essentials. First, he assumes that Euclidean geometry uniquely characterises "real" space, adding that this geometry consists of propositions which are synthetic, thereby extending our knowledge of this space. Second, he insists that this system, qua pure constructions, is applicable within Newtonian physics. Modern mathematicians, puzzled by the purely logical problems posed by the apparent interdependence of the axioms of this geometry, proceed to construct alternative axiom systems based upon this same geometry—so closely based upon it, as to render the axioms and definitions of these alternative systems, Euclidean and non-Euclidean, capable of precise reformulation in terms of any of the others. That is, the propositions of any of these alternative systems are "inter-translatable". [Cf. Poincaré, , p.41] Subsequently, one of these systems—Riemannian—is found to "apply", when interpreted, to the physical world; it is then incorporated into an entire system of physics. This system nevertheless permits Euclidean geometry and Newtonian physics to remain "valid" for special cases specified within the new system—generally, for terrestrial purposes.

It is as if Euclidean geometry has a unique status after all, for it is in terms of it that non-Euclidean systems are constructed; and all the propositions of the latter are formally translatable into the propositions of the former. Indeed, this fact of inter-translatability establishes that as interpreted formal systems these alternative geometries are not logically incompatible. But it follows from this that no empirical considerations could decide between them qua formal systems. Only when they are interpreted and incorporated into physical theories can such considerations be relevant. However, we are then confronted with a choice of *physical* systems, and it is possible (as Poincaré indicates) to retain Euclidean geometry, provided one is prepared to accept more complex physical laws. [Poincaré, *ibid.*]

CHAPTER TWO – KANT'S THEORY OF SPACE AND TIME

So the employment of a pure geometry inside physics need not raise profound problems of "application", since when such a formal system is incorporated into a theory of physics it becomes an interpreted system, and the "objects" that the geometry is concerned with are physical objects. For Kant, the problem of application appears to be avoided, since the act of construction which generates mathematical objects also applies to the objects of science. What we have is not so much a ready-made mathematics applicable to physics, but one fundamental faculty active in both.

Much has been made by other philosophers of Kant's assumption that Euclidean geometry has a unique position as the science of perceptual space. The claim that Kant makes is not that he has proved that there can be only one geometry which describes space; but that in so far as our human experience has been adequately analysed by the critical philosophy, then space, to be given to us as we are, and for our experience as such to be possible at all, must be Euclidean. As early as 1747 Kant had expressed the idea that our Euclidean space is only a particular instance of a more general idea, and suggested, intriguingly, that "...a science of all these possible aspects of space would doubtless be the highest form or geometry which the human mind could conceive". [In Vasiliev, p.103] Kant's motives for playing with this idea—almost, by implication at least, the possibility of n-dimensional geometries, where $n \neq 3$—was his desire to find a philosophical proof of the three-dimensionality of perceptual and physical space, based upon the inverse square law of gravitational attraction. [Cf. Lucas, p.243]

There is no reason to think that Kant abandoned these ideas when the critical philosophy reached maturity. We have now seen that the idea of alternative geometries may be implied by Kant's mature philosophy of mathematics. Kant insists that the axioms of Euclidean geometry are synthetic, and the fact that the denial of one of these axioms leads to no inconsistency would appear to establish this claim. In addition, Kant repeatedly insists that if our finite minds were replaced by other, constitutionally different finite minds, then the nature of experience would be spatio-temporally different. To paraphrase Kant, it is only from the human standpoint that we can speak of (Euclidean) space as the space of perception.

Whether Kant was correct or not in attributing a specific metric to perceptual space, (that is, in moving from space as form of intuition, "spatiality", to space as formal intuition, "determinate" space), he cannot justly be chided for having disallowed those very logical possibilities—now realised—of alternative geometries, which do in fact form the background against which the transcendental justification of his own theory is so sharply differentiated. Had this really been Kant's position, then the very existence of alternative geometries would have compelled a complete abandonment of his theories. Kemp Smith makes the erroneous assertion that although Kant was willing to recognise that the forms of intuition possessed by other finite minds might be different from ours, he does not mean to assert the possibility of non-Euclidean spaces. Kemp Smith says, baldly, that for Kant space, to be space at all, must be Euclidean. [Kemp Smith, p.117] This is Kant's position—but only with important qualifications. In a real sense what Kemp Smith asserts does represent Kant's view—but not as a simple presupposition, but as a consequence of Kant's attempt to prove the synthetic a priori character of geometrical propositions. We need to inject into Kemp Smith's assertion the essential, almost "anthropological" element, referring it always to the context of a possible experience.

The Kantian thesis is far from uncontroversial, but it is not obviously contradictory. Put simply, his general thesis is that many spaces are possible, and logically consistent geometries incompatible one with another are possible, and free from contradiction. Yet the only space of which we, as finite human beings with this particular constitution, can have experience, is correctly described (and proscribed) by the propositions of Euclidean geometry. Indeed, because of the nature of space and time considered from within this human framework, this experience would simply not be experience at all, without these pure

CHAPTER TWO – KANT'S THEORY OF SPACE AND TIME

forms of intuition which serve as conditions for the possibility of that very experience.

Certainly Kant is convinced that space is given to us as being definitely and irrevocably Euclidean. And there is no doubt either that this constitutes one of the major points of conflict between Kant and many of his successors. Perceptual space—the space of any and every possible experience—could not be "re-constructed" such that non-Euclidean geometry could provide the formal basis for intuitive, or particular constructions in space, from which synthetic judgements known a priori could follow.

In the *Prolegomena*, Kant presents what seems to me to be the wholly incredible thesis that not only must space be Euclidean, but that objects *in* space should be also. He says that the propositions of geometry are "determinations" that hold "...necessarily of space and hence also of everything that may be encountered in space". He goes on:

> ...all outer objects of the world of our senses must necessarily agree in all exactitude with the propositions of geometry, because it is sensibility itself that first makes these objects possible as mere appearances, by its form of outer intuition (space) with which the geometer is concerned. [Prol. p.43-44]

Now although we might agree with Kant that the forms of intuition make appearances possible, it hardly seems to follow from this that those appearances must themselves be "Euclidean", whatever that might be taken to mean. Certainly the "structural" form that the appearances (objects) may take will be proscribed by the nature of the space they are in: but to suggest that objects *in* space must necessarily share the geometrical properties of that space appears to be nonsense. What Kant is entitled to say is that the relation between intuition and sensibility is such as to guarantee the possibility of a priori constructions from which the body of synthetic propositions systematised in Euclidean geometry is constituted. Kemp Smith says that "...even—as the modern geometer maintains—should our space be proved, upon analytic and empirical investigation, to be Euclidean in character, other possibilities will remain open for speculative thought". [Kemp Smith, p.119] We have seen that this is in no way damaging to Kant, since it—more or less—represents Kant's position.

I have already drawn attention to the difficulty of maintaining the idea that any system of pure geometry could be either confirmed or denied by experience. Both Euclidean and Riemannian figures are, as such, idealisations: they are not objects on which we can perform measuring operations. Kant's view is inadequate not because it says that the space of experience conforms to the laws of Euclidean geometry. This may well be false—I certainly think it is—but it is in the nature of the case that it could not, as a claim, be empirically refuted. Kant's view is inadequate because of his familiar move from what is the case "anthropologically", to what must be the case. In other words, the very best that we might hope for is a transcendental deduction of space as *formal* intuition, from the nature of space as form *of* intuition. While this might guarantee the uniqueness of this formal system for perceptual space, it could not justify any claims as to the necessity of this structural space for other finite beings. From the bare facts of spatiality, we cannot justify any assertions about the necessity of any metric or determinate space outside our own case.

What is certain is that many modern thinkers have denied Kant's contention that pure intuition is necessary for mathematics. However, in the case of both Hilbert and Brouwer there is an explicit appeal to the necessity of intuition—time, in Brouwer's case, both space and time in Hilbert's. Both appeal to Kant for justification.

The pure intuition of time forms the philosophical basis for Brouwer's theory of number and arithmetic, although he rejects the notion that spatial intuition is also necessary. Hilbert, on the other hand, accepts both forms of intuition from the Aesthetic, and in addition appeals to the Dialectic to make room for the infinite as an Idea of Reason. Brouwer's intuitionism admits only those concepts that can be actually constructed; Hilbert's formalist programme admits the existence of mathematical objects provided they are free of contradiction within the system. If the Kantian thesis is taken as the idea that mathematics is "intuitive", then mathematics will be limited to objects capable of construction. The situation now is that mathematics since Kant has pursued the highest degree of abstractness and generality: it hardly seems to concern itself with space as Kant understood it. Yet even if we accept that logic, mathematics, and mathematical theory have, to a large extent merged, we may still ask if, in addition to the analytical method, mathematics has again been reduced to a vast system of tautologies. The failure of logicism in the last century confirms Kant's own anti-logicist insight, and his conception of construction has been accepted afresh by Brouwer as part of his attempt to avoid formal inconsistencies which arise as a result of the purely analytical methods which eschew intuition in mathematics as an irrelevance. Intuition, as I have tried to indicate, need not—even for Kant—imply a necessary link with sensibility. We can therefore say, with justice, that both Hilbert and Brouwer were Kantians, in spite of their wholly different results.

Certainly the whole programme of a priori construction of figures in space can be seen as dispensable, although it remains heuristically useful. Do we in fact need to appeal to *any* extra-geometrical considerations to justify a geometry? In what sense do the concepts of a geometry require an epistemological or metaphysical dimension in order to have unambiguous meaning?

What we have then, is not only a problem as to the meaning to be given to the question of whether Euclidean geometry is the science of real space. There is, in addition, the question whether it is necessary to go beyond the internal procedures of geometry qua formal system, in order to realise its complete and unambiguous meaning. As I have indicated, there is for Kant no real problem of applied geometry, since the same intuitive, constructive procedures underlie both the object qua geometrical object, and the object qua object of experience, given to us by means of the sensible apparatus whose pure forms, space and time, first make this object possible.

Many discussions in the foundations of mathematics concern the Kantian themes of the nature of infinity, and the decisions as to what constraints a mathematician is prepared to accept. Kant's answer is clear: possible experience, and a priori construction. If we follow Hilbert we can also allow the actual infinite which, although incapable of intuitive construction, can be accepted as an Idea of Reason, providing it with operational meaning. But Kant's conception of infinity is wider than the mathematical notion (as we shall see below), and its philosophical importance would remain even if it seemed inadequate to cope with purely formal problems of mathematical analysis.

2.7 Incongruent Counterparts and the Intuitive Nature of Space

As we have seen, Kant's views on space and time did not remain completely fixed, reflecting as they did his various preoccupations with Newtonianism and Leibnizian relationalism, before reaching maturity in the doctrine of transcendental idealism. His discovery of the synthetic a priori nature of geometrical propositions made some radical transformation of these earlier views necessary, although he had recognised their inadequacies before the critical synthesis had been accomplished. The conflicting pulls of Newton and Leibniz—neither of whose theories were ever totally abandoned—and the concern with the false dichotomy of logic and experience, each played their part in shaping the critical theory of space and time as presuppositional forms. The history of Kant's thinking on space in particular, is highlighted by the argument from incongruent counterparts, which seemed to occupy Kant as a problem (though not always as a paradox) throughout the vicissitudes of his thinking on space. More importantly, the argument is, at the very least, a supplementary proof of the a priori and intuitive nature of space.

The argument first appears in the article entitled "Concerning the Ultimate Ground for the Differentiation of Regions in Space", which appeared in 1768. In this paper, Kant seems moved to write under the stimulus of the Newton-Leibniz debate. He argues against those

"German philosophers" who deny reality to space: at this stage, it seems, Kant supports Newton. But there is much more to it than this, for even though the article seems to be a defence of an absolutist position, it nonetheless contains most of the important insights and considerations which would lead eventually to a rejection of its most explicit conclusion, viz. the existence of Newtonian space, and which developed into the critical view of space as a form of a priori intuition and as transcendentally ideal.

Kant is concerned with the grounds for differentiating "regions" in space. As we have seen, for Newton space as a substantial entity, while for Leibniz, spatial qualities were reducible to the properties of phenomenal aggregates which are not themselves ultimately real. The stimulus of this controversy leads Kant to suggest a justification for an anti-Leibnizian position, and thereby an indirect confirmation of the Newtonian alternative. The argument is not an explicit attempt to prove that space is absolute and not relational; it is more plausibly seen as one that favours the view that space is some kind of reality rather than a merely ideal system of relations.

For Kant, the notion of differentiable regions implies that space is real. He will argue that a full explanation of incongruent counterparts entails reference both to the objects *and* their relations in space, the latter taken as quite distinct from those objects. That the parts of space are only possible through limitations of a given totality of space, is an assertion that Kant makes in all his writings, pre-critical and critical; it is a view that is certainly found in Newton. The positions of the parts of space presuppose a region according to which they are ordered; that is, there is some ordering relation through which all possible positions in space are connected. At this stage it seems clear that the "whole" of which these parts are parts, is absolute space, although Kant nowhere simply assumes this to be the case. A region, Kant says, does not consist in the relations that one thing in space has to the next: this would be "position", or "relation of situation". For Leibniz the region must consist, phenomenally, in relations obtaining between substances by means of "confused" perception; it cannot refer to any external space in which these relations are possible. A region, says Kant, is the relation of a system of positions to absolute space. Now the position of the parts of an object with respect to each other is sufficiently recognised from the object itself. The region to which the order of the parts is directed is related to absolute space: the region, in this sense, corresponds to Newton's "relative" space, and because of the definition of spatial totality in terms of limitations of relative spaces, this region "belongs" to absolute space. It is Kant's intention to provide grounds for the assertion of the independent reality of this system of all possible regions, i.e. absolute space. Any such demonstration would be impossible, Kant says—challenging Leibniz's attempt to base a refutation of Newtonian space on the principle of sufficient reason etc.—if appeal is made merely to abstract metaphysical principles.

Since space has three dimensions, three surfaces can be conceived in physical space. We have knowledge of external things, Kant says, only by means of relations of sensuous impressions to ourselves: and from the relation of these intersecting surfaces to our body we derive the ultimate foundation for constructing the concept of spatial regions.

Kant is endeavouring to show that the complete principle of determining physical form does not and cannot rest merely on the relation and situation of the parts of objects with respect to each other, but must also rest on their relation to external space. This suggests that Kant never intended to show that the difference between incongruent counterparts depends purely on the relation to space, but requires in addition reference to the shape of the object—that is, to internal characteristics. This relation to absolute space cannot be immediately "perceived", but the physical differences that rest uniquely and alone on this ground can be perceived. Such differences are of course the left- and right-handedness of certain objects.

Consider a left and right hand. In terms of geometrical properties, proportions, situations of parts relative to other parts etc., a complete description of one serves as a complete description of the other. "So much may be sufficient to understand the possibility of completely like and similar and yet incongruent spaces". [P.C. p.42; see also Frey, p.112-113] Now Kant can only mean by "spaces" here the region that a body occupies, or the containing surface of the body. This is important, since Kant's most interesting solution to this problem will concern the relationship holding between a body and the external space in which the body exists. Kant's next point, however, is confusing. He says that the determinations of space are not consequences of the situations of the parts of matter relative to each other; rather, the situations of the parts are consequences of the determinations of space. Yet the parts of an object are supposedly describable without reference to external space; I take this to accord with the Leibnizian claim that these relations are not properly describable as spatial at all. Here, however, Kant is saying that the situations of the parts are somehow to be considered as consequences of the determinations of space. But regions, on Kant's view, can be differentiated only through the connections that *objects* have to absolute space. It is difficult to see what is involved in the assertion that space as such could have "determinations" which were not simply relations to already existing objects, unless one presupposes a Newtonian spatial structure of differentiated points which "exist" prior to and independently of physical objects.

After consideration of the geometrical properties of the two objects—the left and right hand—it is clear that some difference remains unexplained. This difference, Kant insists, must be internal. The surface which encloses one, cannot enclose the other. The fact that this "limiting space" can be applied in one case but not the other suggests that the differences rest on some "inner" principle. It cannot, however, have anything to do with how the parts of the object are connected; firstly, because different regions cannot be explained in this way; secondly, because as far as the arrangement of the parts is concerned, the objects are identical. Kant tells us that in the constitution of bodies real differences are found which are connected with "absolute and original space", since it is only through this space that the relation obtaining between physical things is possible.

At the end of the 1768 article Kant makes some remarks which anticipate the position he would eventually reach in his mature philosophy; they also illuminate some otherwise puzzling remarks in that paper, and the use he subsequently made of the argument in later writings. Absolute space, he concludes, is not an object of external sensation; it is a fundamental concept which makes these sensations themselves possible. This suggests Kant's mature position that takes space as a necessary representation underlying all outer impressions of the senses—that is, as an a priori intuition. Kant says that all knowledge of external things by means of relations of sense impressions must ultimately be related to a perceiving subject. I take this to mean that, without having seen the full implications of the view at this stage, Kant has recognised the impossibility of regarding space as a purely intellectual entity which might be described without reference either to objects, which *differentiate* it, or to perceiving subjects, which essentially *constitute* it. It is only by means of the relations of bodies that we can "perceive" the relations to pure space.

There is a short reference to incongruent counterparts in the *Inaugural Dissertation*. Here, the use of the argument occurs within the wider context of a discussion designed to show that space and time are not rational entities, nor objective ideas, but are in fact "phenomena". Whereas in the earlier article it seemed sometimes that Kant was wavering between Leibnizian rationalism, and Newtonianism seen in relation to a whole system of physics, the *Dissertation* reference firmly rejects both of these in favour of a recognisable subjectivist approach, which would develop into transcendental idealism.

All of the important new principles of the critical position are embryonically present in the *Inaugural Dissertation*. The concept of space is a singular representation and is a pure intuition; that is, space is an a priori particular. The argument from incongruent counterparts supports this position, which I believe is latent and unrecognised in the earlier article.

With opposite-handed objects, the diversity—the "discongruity"—can only be noticed by a certain "act" in pure intuition. In the 1768 article, the solution to the problem was that objects differed in their relations to external space; this position is now abandoned in favour of the far more subtle view that leads to transcendental idealism.

However, the solution offered in the *Dissertation* might be considered on other grounds. In the earlier paper, Kant had suggested a provocative thought-experiment. Imagine that the universe is empty except for a single hand. Is this a left-hand, or a right-hand? Given the conceptual indeterminacy of handedness, it does not seem to make sense to assert that it is either one or the other. But now imagine that a handless body is created. Now it seems that the hand must fit on either the left or the right wrist. But if it fits on the left hand, then it must have been a left hand *before* the introduction of the handless body. Kant thought that this shows that some absolute spatial framework—some kind of "ideal lattice"—must exist independently of objects in space. Kant is arguing that any handed object, when considered apart from the conditions of experience, is indeterminate with respect to whether it is left or right. It seems to be a transcendental idealist point at least in the restricted sense in which this implies a doctrine about meaning. A single hand, alone in the universe, is not an object of a possible experience and as such it is literally meaningless to say that such a hand is determinate with respect to whether it is left or right. These concepts presuppose an orientation in space; since space is a form of representation and not an objective, independently existing entity, such orientations presuppose a perceiving mind. There is a clear Leibnizian legacy here: "left", "right", "large", "small", are treated as relational; they require some given framework in order to be fully intelligible. Kant is adding the subjectivity of space as the standard against which such concepts attain significance. For Leibniz, the relations would be ultimately explicable in terms of the attributes of monads; Kant is trying to show that such an account leaves out something important—hence his assertion that the diversity of the hands can only be *noticed* in an act in pure intuition. Ideas like left and right are merely "conventional", and cannot be given a meaning—a sense—in abstraction from perceiving minds.[16]

The transcendental idealist interpretation of the argument is naturally supported by the use made of it in the *Prolegomena*. There Kant argues against those who take space to be a real quality of things in themselves. This complements the argument of the *Dissertation*, which holds that space is a form of pure intuition. Kant points out that various figures, in spite of complete "inner agreement", show an outer relation which makes it impossible to superpose one figure onto the space of the other; and yet the difference is given as being *inner*. This inner difference, which conceptual understanding cannot define, reveals itself through outer relations in space.

What I think Kant means here is that such a difference shows itself only through the relation it has to an experiencing subject for whom space is the condition of the possibility of any and every object in space. Consider opposite hands again: there are no "inner" differences that can be revealed through merely conceptual analysis. Yet the differences are "inner" so far as the sense tell us. In other words, while the objects are similar when considered conceptually, they are nonetheless *experienced* as being essentially different. Now since we can have no experience of space as an independent, self-subsisting entity, these differences must be grounded in us, through space as representation. Kant's solution in the *Prolegomena* is that since objects are not representations of things as they are in themselves, and as a "pure" understanding would cognise them—in an "intellectual intuition"—but are instead sensible intuitions, or appearances, the possibility of which rests on the relations of certain things unknown in themselves to sensibility: then given all of this, the difference between incongruent counterparts can be made intelligible only by referring them to an act in pure intuition.

---

[16] Leibniz had already pointed this out in his debate with Clarke, through the former's denial of the significance of the idea that east and west might, on a *global* scale, be reversed. (See Leibniz's *Third Paper*, in H.G.Alexander, p. 26.) Such a change would be quite without significance; taking a Leibnizian "verificationist" stand, the "lone" hand is not merely indeterminate with respect to left or right; it is also neither large nor small.

There would seem to be three promising lines through which to approach Kant's problem. First, there is the insight of the original 1768 article in which the solution involves the postulation of something like Newtonian space; second, there is the transcendental idealist solution, suggested in the *Dissertation*, and then developed in the First Critique itself, (although there is no mention of the *problem* there); and third, there is the modern approach, which can be traced to both Leibniz and Kant, and relies upon either abstract geometrical considerations, or a more general appeal to the topological features of objects and their relations to the spaces in which they are embedded.

One philosopher, Graham Nerlich, has claimed that Kant's original ideas were almost entirely correct. Nerlich generalises the problem by referring the argument to "enantiomorphic" objects (from the Greek for "opposite shapes"), such as hands, spirals etc. without bringing in the question whether it is feasible to consider the determinateness of such objects. According to Nerlich, what is at issue is the handedness of such objects, rather than whether it is left- or right-handed. He claims that when Kant introduced the handless body into his thought-experiment, he was aiming to show that the object must have been enantiomorphic, not that the hand must have been either left or right. Nerlich says that whether an object is an enantiomorph or not depends on the nature of the space it is in—thus confirming Kant's original insight. So the problem of incongruent counterparts depends for its resolution on considerations of both the shape of the object and the "shape" of the space.

This is the point I attributed to Kant above. Kant says that a complete principle of determining physical form does not rest merely on the relation and situation of the parts of the object with respect to each other, but also on its relation to external space. Nerlich develops this idea, employing the concepts of "orientable" and "non-orientable" spaces to demonstrate how the relationship between the object, and the kind of space in which it is embedded, determines whether it is enantiomorphic or, as he puts it, "homomorphic". Objects are homomorphic if they indifferent with respect to leftness or rightness; for example, opposite-handed triangles in a one-sided non-orientable manifold such as a Klein bottle can be reversed by transporting them through the space by means of a continuous motion. An orientable space is one in which a global definition of left and right can be given. Orientability corresponds to two-sidedness, and non-orientability to one-sidedness—these are intrinsic topological properties independent of any external space.

Now Nerlich claims that it is senseless to say that a hand is neither enantiomorphic nor homomorphic. This is a direct challenge to a transcendental idealist interpretation of the argument which states, in effect, that the question of whether the lone hand is either left or right *is* meaningless, before this is taken in the context of possible experience. Nerlich's point is more difficult. It is one thing to claim that the question as to which hand it is must be meaningless, but it is quite another to say that it is meaningless to assert that it is handed as such. This brings us to topology, and some modern solutions to Kant's problem based upon it.

There is one way of dealing with Kant's original problem which seems both elegant and compelling, and yet it leads the debate into a thicket of epistemological difficulties. The argument can be put like this. Kant was puzzled by how it is that two objects can be alike in essential conceptual characteristics, and yet manifest such obvious dissimilarity. However, if we take the problem down one spatial dimension, the "paradox" admits of a clear and intuitively satisfactory solution.

Consider two opposite-handed shapes in two dimensions; consider further how any two-dimensional "beings" living in such a "universe" (e.g. "beings" whose space was confined to the surface of a sphere) would react to such objects. They would be as puzzled, so the argument runs, by these shapes as we are by enantiomorphic objects in three dimensions. Yet we can easily see that their puzzlement is due simply to the limitations of (their) spatial imagination. By rotating one of these two-dimensional objects in our "extra" spatial dimension, we can superpose this onto the space of the other. The two-dimensional beings would have no way of visualising the process that had occurred—at least not as a continuous transformation of the object through three dimensions. But if—and here the analogy becomes a fairy-tale—if they were sufficiently clever mathematicians, they might be able to work out the properties of a space possessing one more dimension than their own, and thus account for the existence of incongruent counterparts in two-dimensional space. Thus, it is easy to see why we are so puzzled by such object in three dimensions: the "solution" is to "rotate" one of them through a space of four dimensions; we could then superpose it onto the space of its counterpart.

Engaging though this is, such an analogical solution seems to me to raise as many problems as it is designed to solve. The first point I would make against such a view is quite general, and is aimed at all such extrapolations from concepts amenable to mathematical manipulation—n-dimensional spaces, for instance, where $n\neq3$—to real possibilities. In other words, the move from what is logically possible to what is really possible, where the latter concerns those things that could be part of a possible experience. (As I shall point out below, the lasting interest in Kant's problem is due to its being in essence a problem on the *phenomenal* level.) There are not, nor could there be, "beings" of two-dimensions who, in addition to having two-dimensional sense organs, are also capable of perceiving two-dimensional plane figures and have two-dimensional brains that reason mathematically. A further point is less frivolous, but is implied by the first; possible worlds of one, two, or $3+n$ dimensions are abstractions—not real, but ideal. We should no more think of "hypercubes" etc. as real entities, than we should hypostatise points, lines, and surfaces in Euclidean space. Such objects "exist" only in so far as there are mathematical operations and relations incorporating them.

There is no compelling reason to believe in a physical space of four dimensions. The argument therefore remains an analogy which theoretically accounts for two-dimensional incongruent counterparts, but which cannot simply be extended to account for the problem in higher dimensions. The problem in three dimensions concerns the application of abstract geometrical principles to a particular spatio-temporal phenomenon: the analogical argument concerns the application of one abstraction to another.

The purely theoretical nature of this analogical solution may in addition be taken as confirmation of the necessity of the three-dimensional character of perceptual space. We can say that the perceptual space of any being will have the same number of dimensions as opposite-handed objects which cannot be superposed by continuous transformations of one of the objects. Thus our perceptual space has three dimensions, since there do not exist for us incongruent counterparts of two dimensions. It remains logically possible that our three-dimensional space is the "surface" for some $3+n$ dimensional being; but our perceptual apparatus does not allow us to witness the "rotation" of a three-dimensional solid in any higher space as a continuous motion.

There is another consideration that casts doubt on the plausibility of the two-dimensional analogue. The latter suggests that handedness is determined by some putative relation to a higher-dimensional space. But there are good reasons for thinking, as Nerlich indicates, that it depends on the way in which the object is entered into its *own* space. Consider a left-handed shape, in two dimensions, living on the curious surface of a Möbius Band—a non-orientable space having only one side. Such a shape in this case is homomorphic, since transporting around the space will turn it into a right hand. However, unless our three-dimensional space can be plausibly regarded as similarly non-orientable, the equivalent process in three dimensions could not be carried through.

The two ways of resolving the problem by analogies with other spaces leave this alternative: either enantiomorphic objects are, after all, really all homomorphic—because our space is a non-orientable manifold; or we just have to believe in a real fourth spatial dimension. In either case, the original paradox dissolves, but the mysteries deepen in other areas. It has been suggested, for example, that if some of the laws of physics are not left-right indifferent, then it should be possible to say of some object that it is left- or right-handed, even if it was—like Kant's hand—alone in the universe.[17] Since time is one-dimensional, there can be no temporal mirror-images, in the same way that there can be no mirror-images for lines in two-dimensional space. But there is another consideration: space has no preferred direction; if it had, then left- and right-handedness could be uniquely determined, rather in the way that a mathematician deals with it in terms of even and odd permutations, except that in our example the attribution of direction would be non-arbitrary. What Kant has highlighted is that two opposite-handed objects can be *seen* to be such, but not defined. The direction of time always gives us a unique difference between before and after. The absence of an intrinsic direction for space presents us with the problem of mirror-images, which can be directly intuited, but defined only by an arbitrary act of pointing that presupposes this intuition.

The three attempts to resolve the problem—that is, by reference to external space; by appeal to the nature of space as representation; and through topological analysis, all assist our understanding of the original problem, operating as they do on different levels of philosophical abstraction.

Nerlich's solution—appealing to the idea that space itself has a structure—has antecedents in the work of Hermann Weyl. The congruent mappings of space form that group of transformations called the Euclidean group of motions. Once this group is known, congruent volumes may be defined as portions of space that can be carried into each other by continuous transformations. Weyl says that the facts suggest an interpretation according to which the group of Euclidean motions of congruent mappings expresses an intrinsic structure "...stamped by space on all its objects". However, as Weyl points out, the requirement of continuity eliminates transformations of signature "minus": "...a rigid body could go over into its mirror image only by a discontinuous jump.

---

[17] The non-conservation of parity suggests a way of uniquely determining handedness. Since this is impossible in a non-orientable space, physical space cannot be non-orientable. It is possible moreover to consider temporal equivalents of incongruent counterparts; more accurately, spatio-temporal equivalents, e.g. as space-time vectors in some theoretically possible space-time systems. See J.Earman, (1971).

Kant finds the clue to the riddle of left and right in transcendental idealism: the mathematician sees behind it the combinatorial fact of the distinction of even and odd permutations". [Weyl, p.84.] Yet does this not presuppose an independent definition of even and odd? The mathematician, quite legitimately, redefines left and right as minus and plus; he substitutes one convention for another. But unless we had a prior intuitive understanding of the difference between mirror images—unless, that is, we could recognise them as different—we have no basis on which to label them minus and plus. Weyl seems to concede this himself when he remarks that "...to the scientific mind there is no inner difference between left and right. It requires an arbitrary act of choice". He even asks us to consider a thought-experiment like Kant's. Consider, Weyl says, God's creative act; had God, rather than making first a left and then a right hand, begun by making a right one, then He would have changed the plan of the universe not in the first, but in the second act, by bringing forth a hand which was equally rather than oppositely oriented to the first. This seems to me to be part of Kant's central insight in the 1768 article, even if he did not fully recognise it himself.

I remarked above that Kant's original article was concerned not only with offering indirect support to Newton's view of space, but was also a direct challenge to Leibniz and, in particular, to the principle of the identity of indiscernibles. Just how would Leibniz have explained the problem of incongruent counterparts? To some extent, of course, the previous discussion has indicated how a relational view of space might assist in our understanding the issues involved. But there is a deeper problem for Leibniz which operates on and between two levels of Leibniz's system, viz. phenomenal aggregates and ultimately real substances.

The identity of indiscernibles tells us that no two individuals can be exactly alike in all their predicates, and differ solely in respect of spatio-temporal location. This applies to substances—complete, individual monadic histories. The distinguishability of substances is referrable,

CHAPTER TWO – KANT'S THEORY OF SPACE AND TIME

ultimately, to the infinite mind of God, who knows the complete notions of all substances.

Now why should this present a special problem for Leibniz in this context? After all, the difficulty in relation to incongruent counterparts seems, on the face of it, irrelevant here, since it is obvious that we are confronted with two *different* objects—and this is just the problem. But the point is that when we consider the complete notions of two such objects, we have seen above that there is difficulty in accepting that there is an inner difference; they are alike in all essential, differentiable properties. There is no internal property possessed by one which is not possessed by the other. We now see that on Leibniz's principle they should be the *same* object. He cannot differentiate them by reference to relations of external space, because there is no such external space for the objects to be related to in a way that would guarantee their uniqueness. Spatial relations are consequences—phenomenal consequences—of the perceptions of non-spatial monads. There is an obvious difference between two objects that are counterparts, and the principle of the identity of indiscernibles seems, superficially, untouched, since this is precisely what it asserts, viz. that no two objects are incapable of being differentiated by means of internal properties. Yet the principle operates on the level of ultimate reality: the "predicate in notion" principle seems to entail that there must be a predicate possessed by one of the objects which is absent from the complete notion of the other. But Kant has pointed out the difficulty of explicating incongruent counterparts by means of what he would call "merely conceptual" characteristics. For God, who requires no spatial intuition, there can be no difference between left and right. The difference is due either to the relations that objects have to other objects: which would commit Leibniz to the view that the difference is merely phenomenal—but in what could it be well-founded?; or Leibniz must concede that the difference is due to some property of space somehow "imposing" its structure on objects, as Weyl put it. This suggests that space, qua determinate system of metric

117

relations, is "conventional" or ideal for Leibniz, and is "imposed" on phenomenal aggregates.

Leibniz would probably have been much interested in modern views which suggest that there may be physical laws that are not left-right indifferent. Yet as I have indicated, left- and right-handedness as a philosophical problem is irreducibly phenomenal in character, as Kant thought. For Kant, the problem of incongruent counterparts is that it is puzzling how two objects, with identical descriptions, can yet be intuited as obviously different. For Leibniz, the puzzle is that two objects can have identical descriptions, without being the *same* object.

2.8 Infinity: Reason and Experience

Kant's views on infinity and infinite series as expressed (mainly) in the Transcendental Dialectic, seem often to be regarded either as outmoded and lacking in philosophical interest, or as being simply unsupportable. I shall argue that some, at least, of the criticisms levelled at Kant are misplaced, in as much as they conflate mathematical and non-mathematical notions of infinity—a conflation of which Kant was himself not guilty. In this respect Kant's understanding seems to have been strongly influenced by Leibniz, with whose doctrines the Antinomies (where the idea of infinity is central), considered as typical metaphysical claims, is partly concerned.

The critical philosophy is a response to a series of problems with which Kant had grappled in the pre-critical period. The nature of infinity is one such problem. It receives a transcendental idealist treatment in the Antinomies, (of course), but this should not obscure the fact that Kant could not have offered his solution had he not assumed the reasonableness of his own distinction between the mathematical or "formal", and the non-mathematical or "transcendental" conception of the infinite. This distinction remains valid.

In the *Inaugural Dissertation* Kant writes confidently in favour of the idea that the mathematical conception of infinity is consistent. In a highly instructive footnote to Section 1, he dismisses the arguments made against it by showing that they rest on a defective definition. For example, by thinking of an infinite manifold as an infinite number, an inconsistent result can be achieved by defining number itself in a particular way. But Kant says that if we conceive the mathematical infinite as a quantity which, when related to a measure treated as a unit, constitutes a manifold larger than any number; then, since measurability denotes some relation to the scale of the human intellect, through which it is only possible to reach the definite concept of a manifold by successive addition of units—and the complete concept of which is called number only by carrying out this progression in a finite time; we can see that things which do not accord with the law fixed for some one subject *do not thereby remain inconceivable*.

Kant accepts as a logical possibility that there could be an intellect able to see a manifold "at a glance", so to speak, without the successive application of a measure. Given this, there seems to be no particular reason why Kant could not have accepted, for example, the Cantorian concept of infinity. What Kant objects to is the exclusion of a self-consistent notion because this may lead to inconsistencies when its conditions are made explicit. That is, we define number in such a way as to make the concept "infinite number" self-contradictory. Since Cantor defines number in such a fashion as to entail no internal inconsistencies—in effect by first giving a consistent definition to "infinite number" and then defining finite number in terms of it—there seems to be little here to which Kant need take exception.

This is given additional plausibility in virtue of Kant's doctrine of mathematical definition, which requires only that any such arbitrarily invented concept be a logical possibility. For mathematical definitions, consistency is the formal criterion, though "constructibility" is the ultimate test of mathematical reality. The Cantorian infinite is formally consistent; but it remains moot whether we can talk of its being constructed in intuition, notwithstanding the indispensability of formalisms using symbols such as $\aleph_0$, the "smallest" infinite cardinal. I have offered reasons above for thinking that such may be all that is required for Kant to say that some mathematical object has been "constructed". We might need to modify this position in the light of particular demands made by the idea of infinity. I shall say more below of why I believe that Kant's idea of construction may be more than merely "formal" in this sense: a mathematical symbol may be "merely" formal at just that point when infinities make it impossible to provide anything but indirect proofs.

The distinction between mathematical and non-mathematical infinite totalities is highlighted in Kant's discussion of the part/whole relationship. Although his treatment of this subject is complicated by the framework of transcendental idealism, it can be seen to be, quite generally, of the same kind as the Leibnizian distinction between "resolution into notions", and "division into parts". For Kant, it depends essentially on what the object of the analysis happens to be, as to precisely what kind of analysis is possible and, indeed, appropriate. If the object is given in experience, then its parts exist only ideally and are exhibited as discrete by means of an *indefinite* subdivision that is incapable of actual completion: this corresponds to the idea of the "potential" infinite. On the other hand, if the object is given by means of a mathematical construction the parts can be said to be actually infinite.

Kant writes of two kinds of synthesis: the "regressive" synthesis, which proceeds *in antecedentia;* and the "progressive" synthesis", which proceeds *in consequentia*. For example, given the series "m, n, o, p, q, r", where "o" is the point of departure for analysis, "o" presupposes the series "m, n", but not the series "p, q, r"; "o" is possible only through the first series. This would be the case, for instance, if the series was temporal, but it would not be applicable to the timeless series of natural numbers. (The importance for Kant of the time-conditioned series is emphasised below.) The heart of Kant's repudiation of rational cosmology lies in the proposition that the cosmological ideas deal with the totality of a regressive synthesis taking place in time.

Kant further distinguishes the parts of a series from the parts of an aggregate. The parts of space, for instance, are coexistent and not successive; thus space is an "aggregate" or, as the Aesthetic puts it, it is an "infinite given magnitude". One part of space is not the condition for the possibility of any other part. Though space is not a series, the synthesis of its parts, by means of which it is apprehended, is successive, since it takes place in time. There is also a regressive synthesis as regards the parts of matter. Kant should have seen that the parts of matter also form an aggregate, since these, like the parts of space, exist simultaneously. The absolute totality in this case is "given"—the regressive synthesis is not "demanded by reason" in the same sense. Though Kant did not say this about the parts of matter, he does recognise it in connection with the accidents of a substance which, in so far as they belong to the same substance, are co-ordinated with one another and do not constitute a series.

In Kant's view only mathematicians should regard the *progressus in infinitum* as legitimate. Philosophers, he thinks, in so far as they are concerned with conceptual analysis, refuse to accept this expression as legitimate, and instead employ only the idea of the *progressus in indefinitum*. When the whole is given in empirical intuition, the regress in the series of its inner conditions proceeds *in infinitum;* when a single member only of a series is given, starting from which the regress has to proceed to absolute totality, the regress is indeterminate and is *in indefinitum*. This might lead us to think that for Kant the series of all natural numbers is not an absolute totality—thereby denying the actual infinite—since the complete series is never "given" in Kant's sense. However, what he says of definition and construction in mathematics allows him to say that the series is given through and by definition, a functional law. For Kant, we must remember, only mathematics has true definitions which may exhaust the concept through analysis. That is, as mathematicians we can put whatever we wish to in the complete definition of a concept, provided this entails no formal inconsistencies.

I have already noted that the influence of Leibniz is very marked in the discussion of "antinomial" conflict in the Dialectic. Kant says that we are entitled to assert of a whole which is divisible to infinity, that it is made up of infinitely many parts. All the parts are contained in the intuition of the whole, but the whole division is not so contained, i.e. in the regress itself, through which the series is realised. Of course, in saying that all of the parts are contained in the intuition of the whole, Kant does not mean that we have an intuition of all the parts; we intuit the whole, and the whole is made up, quite trivially, of all of its parts. In other words, the parts make up the whole only ideally. It is the law of division that creates the parts; they are not, as it were, already there as an actually infinite collection awaiting discovery.

These considerations enable Kant to assert that the number of parts in a given appearance is, in itself, neither finite nor infinite. What Kant has recognised is the peculiar logic of infinity: this will turn out to be of the highest importance in our subsequent assessment of transcendental idealism. If we accept the limitation Kant places on possible experience, viz. time determinations, it becomes clear that in this context at least, *infinite* for Kant is not simply the contradictory of finite; for it includes within itself contraries that are not equivalent—for instance, the potential and actual infinite. We may deny that some series is infinite; but we must specify what kind of infinity is thereby implied before the disjunction "finite or infinite" can properly be regarded as exhaustive.

Kant objects to the idea that there could be a *constructive* proof of a concept concerning infinity which was not constrained by time-determinations. An appearance and its parts are first given in and through the regress of the synthesis—a synthesis that is never given in absolute completeness. Since this regress is infinite, all the members of the series at which it arrives are contained in the given whole viewed as an aggregate. The series of the division is not so contained. It is a "successive infinite", and is never whole; it does not exhibit an infinite multiplicity. This denies the actual infinite only with respect to the parts of given wholes, where these are given in experience. It is the process of division—the "generating law"—that is given, not the parts of the division as such; these exist only in and through the subdivision.

According to Kant the Antinomies of Pure Reason arise from two separate but related distinctions that have been confused. First, there is a failure to keep distinct what can be defined mathematically in respect of infinite complexes, as against what can be executed as an actual procedure. And second, there is a confusion that results from applying pure concepts or categories of the understanding to something which is "absolutely unconditioned". Reason, Kant says, is seeking the unconditioned in a serial, regressively continued synthesis of conditions—in other words, it is seeking something which stands outside the series, and yet is the condition for that series. The danger arises in not recognising that this absolutely complete synthesis is only an Idea of Reason, a "regulative idea", and not a constitutive principle. We cannot know in advance if the synthesis of appearances is possible at all. Reason starts with the idea of totality, though what it is seeking is the unconditioned, either of the infinite series or of a part of it. [B436]

There are two ways of understanding the idea of the "unconditioned": a) as consisting of the entire series in which all members are conditioned and the totality alone is unconditioned; this constitutes the infinite regress; or b) the absolute unconditioned is only part of a series, a part to which the other members are subordinated, and which does not itself stand under any other condition. So far as a) is concerned, the series is without limits or beginning, and yet is given in its entirety. But the regress in it is never completed, and is therefore properly deemed only a potential infinite. So far as b) is concerned, there exists a first member, which under the aspect of space and time is seen as the limit of the world, and the beginning of the world respectively.

These are the preliminaries to Kant's central concerns in the Dialectic. Most commentators remark critically on the "definition" of infinity given by Kant in the First Antinomy. Kant says this: "...the infinity of a series consists in the fact that it can never be completed through successive synthesis". The problem here is primarily in the term "consists in" (*besteht*), since this suggests that this is indeed intended as a definition. All the same, such a definition might do well enough for non-mathematical uses of "infinity", which is what Kant seems to have in mind here. In the Antinomies, the concern is with the sensible world *in* time and space—not time and space themselves. Of course, the use of the term "successive" does not limit its applicability to temporal series, in spite of Kant's intention here. The point is that it is the application of Ideas of Reason to the objects of the sensible world that generates antinomial conflict; this is what Kant is especially anxious to highlight.

Treating Ideas of Reason as if they were schematised categories generates paradoxes. so when Kant seeks to define infinity, it is from within the context of the world of appearances. That is why it must contain a temporal restriction, since all meaningful judgements about possible experience have such temporal determinations. If we combine Kant's definition of infinity just given with the remarks on infinite regression, we obtain something like this: an infinite series is a series without limits or beginning; we cannot add up the elements of such a series through successive synthesis. Of course put in this way Kant's assertion becomes obviously true, but not trivial, since the notion of the relation of time-determinations to the world of possible experience becomes an essential part of the transcendental idealist solution of the Antinomies; these focus precisely on what can intelligibly be regarded as knowledge of the world.

At this point, this notion of "world" must be made more precise. We could dispense with the term and simply substitute "nature"; for Kant this signifies the sum total of all objects of possible experience—in other words, all spatio-temporal objects. What leads to nothing but confusion, Kant says, is the identification of this with "universe", the latter meaning the "absolute whole" of all coexisting things. Since the "universe" is not, for Kant, describable in spatio-temporal terms, questions concerning its extension and composition cannot arise. In order to think of the world of experience as a whole, we must have counted all the parts; and therefore an infinite time must have elapsed in the enumeration of all coexisting things. But all that this shows is that *if* the world is infinite, (in the sense that the number of things is actually infinite), then we cannot count all the parts. It does not show that the world could not be infinite. This is enough for Kant's purposes, since his primary concern is with what it is possible for us to *know;* the *an sich* reality is beside the point.

The idealist viewpoint expressed here reminds us of the verificationist criterion of meaningfulness—another legacy from Leibniz. It is to just such a criterion of what it makes sense for us to *say,* and what by implication it is reasonable for us to think through judgements and hence to *know,* that Kant is appealing here. Although it does not follow that because a certain operation cannot in principle be carried out that the underlying assumptions must be strictly without meaning, this is certainly the tenor of Kant's argument. Any given totality must be considered as an example of a converging series with respect to its parts. What Kant is considering, with some plausibility, is the impossibility of regarding the "world" as given in this way. That is, only given totalities can be considered to possess an (ideally) infinite number of parts; any object not so given can only be said to contain a potentially infinite number. An infinite aggregate of actual things cannot be viewed as a given whole, nor as simultaneously given.

The thesis and antithesis of the First Antinomy can be summarised from within this discussion as follows. The thesis denies the infinitude of states in time and space, and appeals to the impossibility of making the very idea of infinity intelligible. The antithesis implicitly accepts the inherent meaningfulness of the concept of infinity, basing its claims on the idea of conditions qua causes. Kant says that the true, transcendental concept of infinitude is that the successive addition of units required for the enumeration of a quantum can never be completed. "This quantum therefore contains a quantity (of given units) which is greater than any number—which is the mathematical concept of the infinite". [B461 note.] Kant objects to the idea of the successive addition of an infinite number of units—a temporal process—being completed, while accepting that such a number of units actually exists in the object.

It could be objected against this idea that it is impossible to successively count an infinite numbers of units, that it is true but trivial. Yet in spite of its obviousness, this does seem to be Kant's claim. If he is correct regarding the spatio-temporal constraints of knowledge, then any assertions we make and of which we claim knowledge, must be no better than "pseudo-rational", if they are not restricted in this way to spatio-temporal determinants. Any claims not constrained in this way are either "fictions" or Ideas of Reason. Kant's empirical realism—in the ascendancy here—involves methodological finitism. The completion of the synthesis of an infinite number of units is set as a task, even though the temporal restriction tells us that any such completion is a priori impossible. Yet to stop at any point in the synthesis would involve dogmatic finitism, which Kant rejects.

With respect to space it is impossible, says Kant. to "think a totality", either prior to the synthesis of parts, or by means of it. "For the concept of totality is in this case itself the representation of a completed synthesis of the parts". [B461] Since this completion is impossible, then so is the concept of it. Here, as elsewhere in the Antinomies, Kant argues from the "psychological" impossibility of completing the synthesis, to the impossibility of the concept as such. (Once again, there is a distinction— never far from the surface in Kant—between "logical" and "real" possibility; except that here, logical possibility seems to have collapsed into psychological possibility. The additional complication here is, of course, the logic of infinity.) Once again it is essential to place Kant's remarks within the context of the whole idea of a transcendental philosophy, with its key notion of possible experience. Concepts without intuitions are empty: since no possible intuition could conceivably correspond to the concept of completing the successive synthesis of an infinite totality, the concept can have no application. At best, it is an Idea of Reason; and it must not be confused with a schematised category which, thanks to its link with time, has application to the objects of possible experience.

CHAPTER TWO – KANT'S THEORY OF SPACE AND TIME

In his comments on the antithesis Kant says that the proof depends on the impossibility of conceiving a limit to either space or time. The thesis confuses the "sensible" with the "intelligible" world—of which we know nothing—while the antithesis considers only the *"mundus phenomenon"* and its magnitude, and is thereby limited by the conditions of sensibility. If the sensible world is limited, it must necessarily lie in the infinite void. If that void, and thus space in general as a condition for the possibility of appearances, be "set aside", the entire sensible world "vanishes". This argument presupposes the results of the Aesthetic, of course, and also seems to turn on the ambiguity of the idea of "space in general"; this might suggest either a Newtonian absolute space—a "non-entity"—or the infinite given magnitude of space as a priori intuition.[18]

Kant's remarks concerning limits to space and time raises the question of the nature of limits and boundaries, which Kant had so carefully distinguished in the *Prolegomena*. Boundaries, *(Grenzen)*, with respect to extended objects, presuppose an enclosing space beyond the place of the object; it is thus a "positive" concept. Limits, *(Schranken)*, on the other hand, are negations which affect a quantity in so far as it does not have absolute completeness. In mathematics and science reason recognises limits, but no boundaries. The latter lie beyond present knowledge, even though with respect to appearances knowledge may extend in principle to infinity. Since both mathematics and science are concerned only with appearances, the limits of these sciences are those objects which go beyond possible experience.

It follows from all this that neither metaphysics nor "morals", for example, are legitimate objects for natural science. Any explanation going beyond possible experience ought to be excluded as non-scientific. The transcendental ideas show us the boundaries of the use of pure reason, as well as how these are to be determined. The connection of the known with the unknown may be determined and clarified. If we connect the prohibition to avoid transcendent judgements, with the apparently

---

[18] The distinction between *Void* and *Newtonian space* in relation to Kant has been made by Peter Alexander, 1984, p.6 ff.

contradictory command to proceed to concepts which do not allow of an "immanent" use, it can be seen that both may subsist together, but only on the boundary of all permitted use of Reason. The boundary, as such, belongs both to the field of experience, and to that of pure thought. Given that this discussion of Kant's is in the context of the possible employment of transcendental ideas, and is related to the domains of pure natural science and mathematics, the ideas Kant expresses here are strikingly similar to Hilbert's concept of "adjunction", in the latter's attempt to make the concept of infinite number expressible by means of the accepted propositions of finite mathematics. [See Hilbert, in Benacerraf and Putnam, p.183.]

The fundamental distinction drawn in the Aesthetic between concepts and intuitions is reiterated by the remarks on space made in the "Observation" to the Second Antinomy. Space is not *compositum* but *totum*, since its parts are possible only in the whole, and not the whole through the parts. Furthermore, space is *compositum ideale*, not *reale*. It is an aggregate, not a series. Neither space nor time consist of simple parts. The distinction between what is mathematically real, as against what is real for experience, is the guiding thread running through the discussion at this point, and it is the key to the resolution of the Antinomies.

The Ideas are such that any object "congruent" to them can never be given in any possible experience. It is interesting to note that Kant considered such "illusions" to be possible only in philosophy; mathematics would not and could not generate them, since "...in mathematics no false assertions can be concealed and rendered invisible, in as much as the proofs must also proceed under the guidance of pure intuition and by means of a synthesis that is always evident". In other words, for Kant any science should be in a position to demand and expect none but assured answers to all questions in its domain. In mathematics, at least, problems imply solutions. Such an injunction does not necessarily hold in natural science or metaphysics. It is important once again that we connect this with what Kant says about definition and construction in mathematics. What he is denying is the applicability of pure mathematical method to the problems of "rational cosmology". In other words, there are no algorithms in pure natural science or metaphysics; and in so far as these *are* sciences, it is because of their mathematical content. Definitions are the departure point for mathematics, but the hoped-for point of arrival for philosophy. [B759*a*.]

We have seen that Kant's solution of the conflicts of the Antinomies centres on the key notion of possible experience, and what may meaningfully be said of this from within transcendental philosophy. If the magnitude of the world in space is infinite and unlimited, then it is too large for any possible empirical concept; if it is finite and limited, we have a right to ask what determines these limits. The "cosmical" idea is either too large or too small for the empirical regress, and therefore for any possible concept of the understanding. The fault is in the idea, which is too large or too small for that to which it is directed, viz. possible experience. Since the "world" does not exist as a thing in itself, independently of the regressive synthesis of my representations, it exists in itself—in Kant's paradoxical phrase—neither as an infinite whole nor as a finite whole. It exists only in the empirical regress of the series of appearances, and is not encountered as something in itself. If the series is always conditioned, and therefore can never be given as complete, then the world is not in the required sense an "unconditioned whole", and does not exist as such a whole, whether finite or infinite.

These conclusions of the Antinomies must be among the most controversial in the entire critical philosophy. However, it seems to me that some of the criticisms levelled at Kant's argument at this point in the Dialectic are misplaced. The criticisms, broadly speaking, can be classified under five headings: 1) the alleged use of a non sequitur in Kant's move from the "unthinkable" to the "impossible"; 2) the argument from "elapsed time" in the transcendental conception of infinity; 3) the impossibility of enumerating all coexistents; 4) Kant's assumption that there is only one space; and 5) the impossibility, claimed by Kant, of being able to make sense of the notion of "seeing everything at once".

I have remarked earlier that Kant does indeed argue from what we might call "psychological" impossibility to absolute inconceivability. The move is not uncommon in Kant: it is derived in a quite general way from the implications of the transcendental deduction. Similar moves are common in modern epistemology, where they are more usually referred to as "communication arguments". [Lucas, p.44.] If we are not permitted the transition from the question of *fact* to the question of *right*, then for Kant objective knowledge, uncontaminated by sceptical doubts, will forever elude us. The Ideas of Reason, like the categories, are not abstracted from experience; but unlike the (schematised) categories, they are not applicable *to* experience. The Antinomies of Pure Reason arise when this illegitimate move is made.

There is no doubt that Kant's arguments relating to the idea of "elapsed infinities" have been subjected to severe criticism. One philosopher, commenting on the proposition that "the number of stars in the universe is either finite or infinite", wonders how any important philosophical problem could arise from consideration of such a "self-evident tautology". [Benardeté, 1964.] But it would be thoroughly misleading to assume that Kant wishes to deny both of these two apparently contradictory options. In supporting Zeno against his detractors, Kant says in effect that if the conditions for the judgement in question are such that no genuine knowledge could in principle be obtained through them, then we are confronted with a "dialectical" opposition rather than an analytical opposition. One of Kant's examples makes implicit reference to Leibniz's criticisms of Clarke, based upon Leibniz's verificationist principle; "...if the universe comprehends in itself everything that exists, it cannot be either similar or dissimilar to any other thing, because there is no *other* thing, nothing outside it, with which it could be compared". [B531] Unless there is a genuine contrast, certain judgements lack significance, because it is impossible to specify conditions under which they are either true of false. If the "universe" is regarded as that which comprehends all possible objects and relations, then such relations have significance only from within this systematic totality of possible experience. The universe as such cannot have predicates attached to it in the same way that can the objects within it.

But in any case, is it—as Benardeté has suggested—"self-evident and tautologous" to say that the number of stars that exists is either finite or infinite? It should be clear that an answer depends on just what, precisely, is meant by "infinite" here. The disjunction "finite or infinite" is neither univocal nor exhaustive. If "infinite" here means *potentially* infinite, Kant would find no difficulty accepting it. For Kant it is only the *actual infinite*—an actually completed or completable totality—that is excluded when a proposition concerns some reference to appearances. But the actual infinite is further differentiable into the denumerable, and non-denumerable infinity. The proposition "the number of stars that exists is either finite, or potentially infinite, or denumerably infinite, or non-denumerably infinite", no longer has the appearance of self-evidence. This is not to deny that the disjunction does not exhaust the possibilities; it is simply that we suddenly have the idealised mathematical infinite putatively applied to the enumeration of an empirical collection without further argument. But the Cantorian conception of the actual infinite (for example) can hardly be said to univocally apply to such collections. This would involve treating spatio-temporal particulars as if they were elements of an idealised series.

Consider the natural number series, and one of its proper sub-sets, e.g. the series of squares. Following Cantor, these two series can be placed in one-to-one correspondence—they have the same cardinal number. Now consider the empirical collection of all the stars that exist; a proper sub-set of this collection consists of elements which are, unlike numbers, spatio-temporal particulars. (Numbers are not spatio-temporal particulars, although of course numerals are; the latter, as Kant recognised, are "symbolic instantiations" of the former.) If we try to place (if only in imagination) this sub-set in one-to-one correspondence with the whole set, (i.e. if we assume that we can move them around like the beads of an abacus), we must, so to speak, leave "gaps" in the original set. If it is objected—as it surely must be—that we are really only concerned with the question "How many?", and that all we really need to do is attach, say, a numbered label to all the particulars; then in this case we are treating the set as if it was already a series of numbers—which is exactly the point at issue. We would be considering an empirical collection as an ideal collection.

Of course, this is what we do all of the time in using arithmetic to count finite collections of empirical objects. The problems arise—as Kant well understood—only when this "harmless" procedure is extrapolated to the numbers and operations of "infinite arithmetic". In this respect, Kant's transcendental idealist solutions to the problems raised in the Antinomies entails a thoroughgoing empirical realism. He also points out time and again the difficulties accruing when philosophy borrows the mathematicians precise concepts, and applies them in some imprecise way or in some inappropriate context. Kant never tired of saying that the respective methods of mathematics and philosophy were irreducibly different; the application of one method in the domain of the other would lead to nothing but confusion: "...in philosophy the geometrician can by his method build only so many houses of cards, just as in mathematics the employment of a philosophical method results only in mere talk". [B755]

Any discussion of the Kantian "assumption" that there is only one space would take us too far into speculative metaphysics and in any case is not really to the point here. Certainly Kant takes it that any space of which we have knowledge must be related to us in some direct, spatial way. Any wild talk of "parallel" but non-connected spaces would be, almost by definition for Kant, beyond any possible experience. This is part of the claim of the Third Analogy: "All substances, in so far as they can be perceived to coexist in space, are in thoroughgoing reciprocity". [B257] If we can perceive two objects as coexisting, then it must be possible (in principle) for causal influence to pass from one to the other, and from either to a perceiving subject. If the two objects belonged to different spatial systems, not only could they not mutually interact, but they could not be perceived by someone at one and the same time. If the latter were possible, then it must be possible for the objects to interact, which implies that they share a common space.

Finally, Kant nowhere denies that the "perception" of an actually infinite set could not be possible. What he denies is that this is a possibility for us. But "...there could be given an intellect, which might see an (infinite) manifold distinctly at a single glance without the successive application of a measure". [P.C. p.49*n*.]

The transcendental ideas have, for Kant, a proper use within philosophy. They are regulative principles, which have the function of directing the understanding towards a goal upon which the routes marked out by the rules of the understanding, themselves converge. The point of intersection of all such lines is a mere *focus imaginarius*, which itself is not to be found in any experience, since it lies outside all experience. It gives systematic unity to that experience, serving as a guide to development. For example, we make use in empirical science of certain notions which are idealisations that do not pretend in fact to be experiential—the concept of an ideal gas, for instance. Only by postulating such ideal concepts are we able to determine the actual structure of those experiential elements which are given to us.

The Ideas of Reason, as we have seen, are not opposed to the categories, but to the schematised categories. Unlike the latter, the cosmological Ideas cannot be given a transcendental deduction; and if we assume they are objective and constitutive, we generate antinomies. The transcendental Ideas are "schemata" of the regulative principle of the systematic unity of our knowledge of nature. They are not real things, but rather *analoga* of real things. The Ideas are therefore maxims of speculative reason, and as such do not conflict either with one another, or with the schematised categories. (It should be recalled that in Kant's moral philosophy, the "maxim" plays the role of a mediating concept between Reason and inclination.)

Thus regulative Ideas, as principles, are neither true nor false. We can "apply" the cosmological ideas only analogically in order to form some concept of "intelligible" objects which could not, in the nature of the case, come before us in any possible experience. These Ideas are thought only problematically—they are "heuristic fictions". Outside of this, they are mere thought-entities, *(Gedankendinge)*, which cannot be demonstrated, and are not to be employed as hypotheses in order to explain appearances. If they were so regarded, the implication would be irresistible to think of them as capable of verification. As fictions, the Ideas can be seen as auxiliary constructs which are justified, without of course being susceptible of a deduction , in Kant's sense. The Antinomies arise when these Ideas are hypostatised—when *non Entia imaginaria* are mistakenly transformed into *Entia*. This is most evident in the case of spatial and temporal continua, and the putative infinity of space and time. The actual infinite, as Hilbert remarked, is nowhere realised; it has the role only of an Idea of Reason in Kant's sense. It has an immanent and never a transcendent use, but even then only in so far as it unifies the search for empirical knowledge; it is not itself a part of such knowledge.

2.9 Transcendental Idealism

The thesis of transcendental idealism can be summarised in the assertion that we can have knowledge only of appearances, and never of things in themselves. The doctrine involves several different claims which, though interdependent, can profitably be discussed separately. The different elements in the Kantian theory may be put as follows: 1) a recognisably intersubjective theory of space and time; 2) an explanation of geometrical and scientific principles as synthetic a priori judgements; 3) the limits to possible knowledge, given the plausibility of the general principles justified in the First Critique as a whole; 4) the inadequacies of alternative accounts of metaphysical knowledge, as examined in the Dialectic and the Refutation of Idealism; 5) the doctrine of Phenomena and Noumena. It is obvious that 1) to 5) taken together will yield transcendental idealism, but it also seems to me that one is committed to Kantianism in some eviscerated form if one accepts 3) and 4), yet rejects 1) and 2). The

CHAPTER TWO – KANT'S THEORY OF SPACE AND TIME

doctrine of Noumena is difficult to make consistent with the remainder of Kant's principles, even for the most sympathetic interpreter.

We have seen that Kant's philosophic imagination was aroused by the kinds of metaphysical disputes of which the Clarke-Leibniz correspondence is typical. In that dispute it seemed that the same principles—the principle of sufficient reason, for instance—could be appealed to in support of completely opposing positions. Resolving such disputes might be possible in four ways: we could say that such principles are themselves faulty; or that something has gone wrong with the application of the principle to its objects; or, thirdly, that something is not sufficiently clear about that to which the principle is being applied. Kant shows that the last two possibilities are indeed the cause of such disputes, but also that such errors are at the root of all metaphysical disputation, and that a failure to recognise this has been the reason why metaphysics has never developed as a science. The questions to be asked are always the same: first, given any concept, what is its extension?; and second, is the domain of objective knowledge such that some concepts are forever proscribed from having an extension in that domain?

These are really aspects of the same overarching metaphysical problem, and it is for Kant the task of the critical philosophy to provide a definitive answer. Metaphysics has amounted to mere talk, so far as Kant is concerned, simply because the wrong questions have been asked. Transcendental idealism is therefore an answer to a serious and long-standing metaphysical problem. Kant claims to show that when certain concepts are misapplied, antinomial conflict is the inevitable outcome. We must, he says, distinguish categories, which when schematised have application to the objects of experience by means of a temporal link, from the Ideas of Reason, which cannot be so applied, but as regulative principles may guide and systematise our search for empirical knowledge without thereby adding to it. For example, trying to apply a concept to our idea of "universe" will generate paradoxes. Such a concept, through its limitations to our forms of intuition, really has only intelligible application to "Nature", considered as the totality of spatio-temporal objects.

It is worth recalling in this context that Leibniz had attempted to resolve another famous conflict by means of an argument that may well have influenced Kant. The paradoxes of motion associated with Zeno arise, according to Leibniz, because we try to apply an idealised notion—viz. infinite divisibility—to an empirical object or process. A line contains an infinite number of points only in an ideal, purely mathematical sense. We are seriously in error if we think we can apply an ideal notion to the domain of sense-experience in any complete or precise manner. The two realms of mathematics and sense experience are logically disconnected; applying one to the other indiscriminately commits what Kant referred to as the "subreptic fallacy". [P.C. p.81-82; and see 3.5 below.] What Kant points out in the Antinomies is that it is not so much the concepts we apply that are faulty, but that to which they are applied. This is the lynch-pin of the critical philosophy: the notion of the possibility of experience, considered as a non-logical, non-empirical source of knowledge, expressed through judgements that are synthetic and yet a priori. The universe "as a whole" is not an object of a possible experience, and as such we can say nothing positive about it; and we certainly cannot expect to know anything of it through concepts employed outside of the domain within which their intelligibility has been shown to be restricted. The general result of the Transcendental Dialectic has been to demonstrate the futility of metaphysical disputes which have not been undertaken in recognition of the limits of reason.

If we consider the Aesthetic in its negative role, we can see that it is intended as a means of avoiding what Kant regarded as the unfortunate consequences of accepting either the Newtonian view of space and time as absolutes, or the Leibnizian alternative that they are merely ideal. For Kant, these views are not just mistaken in themselves, (for reasons already offered in the Metaphysical Expositions, and for the supplementary considerations of the argument from incongruent counterparts); but in addition they are incapable of properly explaining

CHAPTER TWO – KANT'S THEORY OF SPACE AND TIME

the synthetic a priori nature of geometrical propositions. Elements 1) and 2) are thus reciprocal for transcendental idealism, and cannot be understood, from Kant's point of view, apart from one another. The *quid facti* requires a *quid juris*.

It is commonly levelled at Kant that he argues from the nature of geometrical propositions to the nature of space as a priori intuition. There is no doubt I think that this reflects the genetic aspect of Kant's theory; but logically, Kant at least believes that the status of geometrical propositions is a reason for holding his theory to be true, rather than premises from which his theory of space is deduced. [See Horstmann.1976] Certainly Kant gives reasons for accepting the view of space as a priori intuition which do not depend on the argument from geometry—the argument from incongruent counterparts, for instance. It does seem from the pre-critical writings that Kant first discovered the constructive nature of mathematical reasoning: and this, coupled with his belief in the truth of the propositions of Euclidean geometry, (something neither Newton nor Leibniz thought worth denying), led him to ask some fundamental questions about space, and the science of space.

Although it is true that much of the Critique can be seen as a philosophical underpinning of Newtonian physics, there can be no doubt as to Kant's rejection of Newton's view of space and time. What Kant is rejecting as much as anything else is that this space and time is mind-independent, and the "Copernican Revolution" in philosophy that Kant has wrought requires that the forms of knowledge be active contributions of the mind. As I noted earlier, Kant saw the concepts of absolute space and time as *"Undinge";* he says that "...the mathematical students of nature [the Newtonians] have greatly embarrassed themselves by those very conditions (space and time, eternal, infinite and self-subsistent), when with the understanding they endeavour to go out beyond this field". [B57] Kant is about to demonstrate in the Critique the impossibility of extending our knowledge into the realm of the super-sensible; he is certainly not going to resist the chance of repudiating two such influential super-sensible realities as Newtonian space and time.

The replacement of these notions by his own theory of space and time as non-relational a priori forms of intuition, cannot effectively be achieved before the task of the Aesthetic and the Analytic has been carried through. The refutation of Leibnizian space and time however, is made, in principle, before the critical reconstruction has begun. The First Critique is an answer to the question: How are synthetic judgements, valid a priori, possible? Kant is in no doubt that there are such judgements—geometry is the prime example of a science which employs them. According to Kant the relational theory of space and time does not allow an explanation of this fact, so the relational theory must be rejected. If space and time are concepts rather than intuitions, the synthetic a priori nature of geometrical propositions is inexplicable. If they were a priori concepts, these propositions would be analytic; if they were empirical concepts, the propositions would be empirical. But Kant insists that the propositions of geometry must be "apodeictic" (necessarily true), yet still valid of objects. He is convinced that only his theory of transcendental ideality can account for this dual nature.

Transcendental idealism involves empirical realism. This implies that space and time are objectively valid in respect of all objects that can be given to us in a possible experience. If we abstract from the conditions of knowledge, which is mediated by space and time, then space and time are "nothing" and cannot be ascribed to objects as they are in themselves. Kant is agreeing with both Newton and Leibniz in saying that space and time are not properties of things as such; for Newton they are containers *for* things; for Leibniz confused perceptions *of* things.

It might be assumed that this assertion in respect of space and time would be as much as any modern view could concede. Yet this would overlook the significant connection between any purely pragmatic conception of knowledge and its implicit metaphysical base, which involves, for Kant, the key notion of possible experience. This cannot be properly explained without considering the conditions which make any objective experience possible; and this means accepting that space and time are transcendentally ideal. The minimal claim might be that since all

CHAPTER TWO – KANT'S THEORY OF SPACE AND TIME

knowledge is (for Kant) necessarily mind-involving; and space and time are the conditions which make this knowledge possible; then the ideas of space and time must be mind-involving. This does not exclude, a priori, the possibility that space and time are somehow given with the contents of perception. But this makes it difficult to avoid the conclusion that the propositions of geometry are empirical, when even modern opinion leans towards the view that such propositions are not strictly applicable to the empirical realm at all; that is, there is no precise "fit" between ideal objects and empirical objects. But there is a more obvious objection. If we wish to avoid the conclusion that the propositions of geometry are empirical, and yet leave open the possibility that the contents of perception come with spatio-temporal qualities; (that is, that the form and matter are empirically given); the only alternative remaining would be to assert that nonetheless our ideas of space and time are yet mind-dependent, (trivially true, since only minds can have ideas); and then accept a pre-established harmony between our ideas and the spatiality and temporality of that which is presented to us. Kant wishes to avoid any taint of rationalist metaphysics implied by accepting a principle such as a pre-established harmony. This would be to revert to a pre-critical solution—in fact, it would be no solution at all. It would make the relationship between the mind and its objects ultimately unintelligible, and instead of offering a solution to a problem, would merely christen it, (as Collingwood put it) with a long name.

Transcendental idealism implies that our knowledge of the world essentially involves a transaction between ourselves and external objects where the perceiving subject is actively involved, and is not merely a passive recipient of impressions. Transcendental realism is excluded as an account because of this active component; we have no "intellectual intuition" of objects—that is, an intuition unmediated by space, time, and the categories. We are in no such immediate relation to objects as they "really" are, as is assumed by a realism that is not linked to a transcendental idealism. We can summarise the Kantian position in the truistic-seeming proposition that we can only know things in so far as our knowledge is limited by the conditions imposed upon it by our faculties. But this "truism" is denied by those philosophies that insist we can know "essences" as well as mere existences. Its denial is usually related to the kind of extension of the application of concepts that Kant exposed as illegitimate in the Dialectic. But whether Kant is himself wholly free from this charge may be questioned by examining his own employment of the concept of an unknowable entity, viz. the noumenon.

The very idea of noumena is implied the moment the full-bodied thesis of transcendental idealism is accepted. In particular, if we regard the doctrines of the Aesthetic as sound then, according to Kant, the objective reality of noumena is assured. [A429] If we accept that we are only ever aware of objects as they appear, then we seem prima facie committed to the idea of the object "as it is in itself", distinct from those appearances. We must not assume that Kant is merely drawing attention to a linguistic peculiarity here—"appearances" always being "appearances of" something. He makes both positive and negative claims for the idea which go well beyond the use of language. If we regard an "object" in so far as it is *not* an object of our sensible intuition, i.e., by abstracting from our mode of intuiting it; this yields the negative sense of the term. But if we go further, and consider it as an object of intellectual, non-sensible intuition, thereby assuming the (logical) existence of such a mode of knowledge, then this yields the idea of the noumenon in a positive sense. According to Kant, both ideas are at least free of contradiction, since we have no reason to think that sensibility is the sole mode by which objects can be intuited, at least in principle. *De facto*, there are no human minds capable of such intuition, but this in no way prevents us from forming a legitimate idea of it.

Having isolated the various elements involved in our knowledge of objects, and asserted that these constitute our modes of knowing, Kant believes that it is quite consistent with this to hypothesise some other mode of knowing, not restricted in the manner he has shown human knowledge to be. The noumenon, he thinks, is actually necessary in order to prevent sensible intuition from being extended to things in themselves.

The *concept* of the noumenon is a "limiting" concept, "...the function of which is to curb the pretensions of sensibility". [B311] And yet it is not to be regarded as arbitrary, since it is a necessary adjunct to sensibility considered as limited. (This reference to the noumenon as a "limiting concept" must not be confused with the mathematical notion of a limit. The latter is well defined by a numerical series, for instance, and may be "approached" through successive approximations; it is the *concept* of the noumenon, not the noumenon itself, that is the "limiting" concept; it is an adjunct to sensibility, and does not become known by any number of approximations.)

How is Kant able to assert that the idea of noumena is of a *plurality* of things in themselves? Since we cannot apply categories to this idea, then neither can we apply concepts such as unity, plurality, causality etc. to it. Can we even say that there are noumena? Is there one noumenon for each appearance; or is the realm of the thing in itself sufficient to embrace them all? We are now asking questions relating to an intelligible realm of essences, and are dangerously close to abandoning Kantianism at precisely that point where the critique is supposed to be of most help. Is the very idea of the noumenon un-Kantian and to be dismissed out of hand?

The necessary mind-dependency of knowledge commits us to the view that there must always remain areas of reality that are unknown and inaccessible to us. The existence of alternative conceptual frameworks shows that Kant was correct in asserting that we can know things only as they appear. The fact that there are alternative ways of knowing something (even ways incompatible with the Kantian categories) implies that what is known depends upon the framework by which and through which it is known. All knowledge is knowledge from some perspective or other and by means of some framework or other. I can know an object from different perspectives; for example, I can know my desk as a functional item of furniture, or as a collection of atoms and molecules arranged in a certain way. These alternative views in no way conflict; the first is a commonsense description, the latter an idealisation that replaces the description for certain purposes and does not compete with it. Different levels of description and explanation are (largely) autonomous. Not only is it not possible to say that physical descriptions replace commonsense descriptions; it is also possible to say that within physics (or within science generally), there may be alternative frameworks not reducible to one another. For example, the physical descriptions and laws applied to macroscopic events and objects, (for instance in cosmology) are sometimes disconnected from, and not reducible to, the laws and descriptions applied in microphysics. Such a "two-worlds" view is acceptable not only as applied to the relationship between experience and theory, but is also perfectly plausible intra-theoretically: physics as a theoretical science accommodates separate theories—General Relativity, and Quantum Mechanics, for instance.

So objects can in principle be known through an indefinite number of perspectives. But these perspectives do not *constitute* the object as such; they constitute the object as it appears and is known by some thinking subject or subjects, employing some framework of description or explanation. The object "as it is in itself" exists independently of all such perspectives. But such an object cannot be known, because all knowing is in this sense perspectival—that is, in terms of some explanatory framework or frameworks, which are neither immutable nor unique.

There is nothing here that leads inexorably either to scepticism or Berkeleyian idealism. It is simply a recognition of the limitations of human knowledge, and in no sense a repudiation of such knowledge. We require, as Kant knew, a sceptical method rather than scepticism as such. It is only by recognising that knowledge is limited in this way that the distinction between knowledge and ignorance makes sense.

We could perhaps give up the idea of the noumenon in its positive sense, yet retain the negative sense. Indeed, as soon as one recognises that some areas of knowledge are beyond the reach of any and every empirical enquiry, it seems that one is committed to the view that the objects of scientific and commonsense enquiry must be what they are independently of those enquiries; the idea of the noumenon need include no more than this. Any attempt at a positive characterisation of this transcendental object must assume the possibility of comparing this object, at least in principle, against our knowledge of it. Since this object is supposed to be independent of our knowledge of it, there is no intelligible manner by which such a comparison could be made. Naive realism transcends its own principles if it insists that the world as it is (currently) known is the real world; admitting the limitation of what we know at any one time sets us the task of looking further. We cannot dogmatically stop at any one point, but we cannot guarantee the success of the continuing scientific and philosophic enterprise either.

Chapter Three

ACTS, INTUITIONS, AND CONSTRUCTIONS

3.1. Introduction

In this chapter it is my intention to abstract from the previous discussion a coherent thread of ideas contained in four very different arguments, which when taken together provide the core of an interesting and convincing picture of Kant's transcendental philosophy in general, and his philosophy of mathematics in particular.

These four arguments are as follows: first, the argument of the Schematism; second, Kant's constructivism in the philosophy of mathematics; third, the argument from incongruent counterparts; and fourth, Kant's (neglected) remarks on indirect or "apogogic" proof from the Transcendental Dialectic, and the use made of such proofs in the Antinomies of Pure Reason. A great deal of philosophical time has over the years gone into an analysis and assessment of most of these arguments taken in isolation, but there is little evidence suggesting that philosophers have found it useful to consider them together. I suggest that they are mutually illuminating, even though it is not clear that Kant himself considered them in quite this fashion. What I have to say here is therefore a re-construction; but it is, I believe, firmly anchored in Kant's writings.

First, some preliminaries. There is a curious exegetical link between the argument of the schematism on the one hand, and the argument from incongruent counterparts on the other. I attach no special significance to this link; it merely seems to me to invite comment. What I have in mind is what might be called the "anomalous" nature of the two arguments within Kant's writings. Consider first the schematism: apart from the fact that it is regarded by commentators as being everything from absolutely central to an understanding of the entire critical philosophy, (e.g. Heidegger), to completely superfluous, (e.g. Wolff), it has also been pointed out that the explicit results of the schematism argument seem not to have been used by Kant in quite this way anywhere else in the First Critique. [Pippin, p.125] Perhaps this does indeed require an explanation; unless, that is, we assume simply that all subsequent mention of categories after the schematism chapter must somehow presuppose the results of that chapter—a possibility that does not seem at all plausible.

Next, consider the argument from incongruent counterparts. I take it to be absolutely crucial for a full understanding of Kant's theory of space. [See also Buroker; and Walker, p.48ff.] I believe it is tempting to go so far as to suggest that the argument is Kant's real reason—or at any rate his best reason—for holding that space is an a priori intuition. It would also appear to offer a supplementary philosophical argument for the 3-dimensionality of phenomenal space.

If all of this is correct, why then is there no mention of the argument in the *Critique of Pure Reason* itself? Indeed, even if the argument is not as strong as Kant, or some later commentators, have taken it to be, it can hardly be said to be anything other than supportive of Kant's criticisms of other theories of space and time, even if it does not in fact provide any conclusive evidence for Kant's positive theory. It seems that Kant was bothered by the questions raised by this argument from 1768 onwards; and the argument gets a brief mention in the *Prolegomena*. If, as I shall argue, the metaphysical thesis of transcendental idealism offers a framework within which some at least of these problems admit of resolution, why not mention it somewhere in the First Critique? There is no doubt that its absence is very puzzling, and that it has been an irritant to the most sympathetic and imaginative of commentators. There is the merest hint of the argument in the First Critique, as I shall point out below, but I would not argue very strongly for the idea that it is intended by Kant as some sort of surrogate for the full account. It may be that no convincing argument can be given for its omission, in which case we shall have to be content with the re-constructed argument I offer below that it is, nonetheless, very important, even if Kant, for whatever reason, chose to overlook it.

3.2 Concepts, Intuitions, and the Schematism.

What can we say, then, are the permanently important results of the schematism chapter? I would list them under five headings, as follows: i) schemata are the "bridge" between the intellectual or conceptual, and the sensible; ii) schemata are rules for construction, rather than images; iii) schemata provide the essential (transcendental) link with time; iv) so far as the schematism is concerned, it is the *act* of construction that is important; and v) there is an isomorphism between the a priori truths which belong to the concept, and the identifiable a priori conditions exemplified in the construction.

Kant's idea that there must be some way of bridging the gap between that which is sensible and that which is conceptual is of course highly controversial in itself. I believe that this is a genuine problem that Kant sets himself to solve and that it cannot be collapsed into the "pseudo-problem" of the distinction between having a concept and applying it. Even given this, it is puzzling that Kant should find the problem of justifying the existence of pure concepts so much more intransigent than the problem of applying those concepts, after the fact of their existence has supposedly been established by the Transcendental Deduction. We would be entitled to assume that this is Kant's position, at least, given the space devoted to the Deduction and the Schematism arguments respectively. I therefore disagree with those who see the schematism as more of an "explanatory pause" than a new step in the argument. It has been suggested that the results of the Deduction are in themselves supposed to do the job of restricting the use of pure concepts; but though this may be so, it omits the essential generality of those results. What the transcendental deduction tells us is that the categories apply to objects of experience in general: Kant takes it that there is an additional issue concerning the application of those concepts in determinate instances.

I have said that, for Kant, the schema provides a bridge between the intellectual or conceptual on one side, and the sensible on the other. In a very important sense, this problem is quite central to a very great deal of philosophy. How is the material of experience to be connected to the theories that we hold true of it, since, as Kant points out, these are quite heterogeneous? That there is such a distinction to be made seems to be presupposed even by those philosophers wishing to collapse it—whether the result is some form of empiricism or phenomenalism, or whether it is absolute idealism at the other extreme. The whole tenor of Kant's critical enterprise is to find a way of avoiding the rift between empiricism and rationalism while doing justice to the contributions made to our knowledge of the world by matter and mind respectively. Experience is an irreducible amalgam of the conceptual and the sensible; but is Kant not correct is assuming that the different contributions made by sensibility and understanding do not—indeed, cannot—in themselves, explain their "co-operation" in this experience? In other words, neither intuitions nor concepts can contain any explanation of how they might amalgamate in an experience.

That task must fall to a "meta-theory" of knowledge which, as I see it, is precisely what Kant is aiming to furnish. The lynchpin of that meta-theory is Kant's explanation of the nature and significance of synthetic judgements known a priori: further, the schematism is crucial to our understanding of this synthetic character. [See also Allison, 1981, p.57-83.]

I also reject the view sometimes advanced that the whole idea of a "third thing" mediating between a concept and an intuition is "incoherent". [Gram, 1968] This idea of a third thing which is both universal and particular, intellectual and sensible, must be given up, say its critics, since no "object" can possess such contradictory properties. But it seems to me that, in a clear sense, this is what an *example* can do; that is,

CHAPTER THREE – ACTS, INTUITIONS, AND CONSTRUCTIONS

it is in the nature of the case that an example just is the presentation of something universal in something particular, and that this is central to an understanding of the nature of mathematical constructions.

With this said, it is nonetheless difficult to be unambiguous about just what kind of strategy Kant pursues in the schematism chapter, and just what he takes the "application problem" to be. The two most obvious candidates for the resolution of that problem concern the "judgemental", and the "subsumptive". I will remark on these in turn.

It has been suggested that for Kant it is the syllogistic rather than the judgemental concept of subsumption that is at work in the schematism. [See Allison, 1981, p.64.] Although it seems unlikely that Kant would wish to construe the application of categories to intuitions as a kind of syllogistic reasoning, Allison at least concedes that this analogy underscores the particular problem which occurs when the attempt is made to understand such application. What Kant requires is an analogue of the condition of the rule—the middle term of the syllogism—under which appearances can be subsumed. And this analogue is the schema. The central point is the idea that the schema is itself to be construed as a pure intuition. [See Gram, *ibid.*] Although I think this is mistaken, the reasons given for it are interesting in themselves and deepen our understanding of the problem.

We can certainly agree that Kant's various characterisations of the schematism are confusing and unnecessarily complicated. We already have the notion that it is some "third thing" between concepts and intuitions that yields knowledge of determinate objects. [A138] In the schematism chapter we can also find the following: that schemata are transcendental determinations of time; [A139]; that schemata are formal conditions of sensibility to which the Understanding is restricted; [A140]; that a schema is a representation of a universal procedure of imagination in providing an image for a concept; [A140]; that schemata are to be understood as the function of pure synthesis, to which the category gives expression; [A142]; that they are "significance functions" for concepts; [A146]; that they are a priori determinations of time in accordance with rules; [A145]; and finally, that the schema is the phenomenon, or sensible concept of an object, in agreement with a category; [A146].

To defend the thesis that schemata are pure intuitions requires us to say that either all of these characterisations are compatible with such an idea, or that they are all somehow reducible to it; or, more drastically, that one or more of these characterisations can simply be ignored as an aberration on Kant's part. It does seem that, *prima facie*, only Kant's idea of schemata as formal conditions of sensibility will strictly accord with this particular thesis. It would certainly appear that the pure intuition thesis is completely out of sorts with the notion that schemata are representations of universal procedures of the imagination in providing images for concepts. We would therefore need to argue a case for construing schemata as pure intuitions on the grounds that the thesis is compatible with Kant's affirmation that they are a) formal and pure conditions of sensibility; and b) that schemata are "significance functions" for concepts. It is possible to find support for this in Kant: it is true that Kant himself seems to identify schemata and pure intuitions in both the Second and Third Critiques. Now although he does indeed appear to suggest this in the Second Critique, the passage is hardly conclusive evidence, especially in the light of what Kant says shortly afterwards in the same work, and in light of what would seem to be, in the case of the relevant passage of the Third Critique, an example of Kant's inconsistency in the use of key terms. In the Second Critique, Kant says this:

> The judgement of the pure practical reason is subject to the same difficulties as that of the pure theoretical reason. The latter, however, had means at hand of escaping from these difficulties, because, in regard to the theoretical employment, intuitions were required to which pure concepts of the understanding could be applied, and such intuitions...can be given a priori, and therefore, as far as regards the union of the manifold in them, conforming to the pure a priori concepts of the understanding as schemata. [C.Pr.R., p.160]

CHAPTER THREE – ACTS, INTUITIONS, AND CONSTRUCTIONS

In the Third Critique, Kant writes that all concepts, whether pure or empirical, require verification: empirical concepts are verified by "examples"; pure concepts are verified by schemata. In the Second Critique, Kant had warned of the error of turning what serves only as a "symbol" into a schema: now, in the Third Critique, he seems to have conflated their respective roles himself. [See C.Pr.R., p.162.] But in any event, neither of these passages seem clear enough to establish or confirm the schemata as pure intuitions thesis.

Other reasons are offered in support of this idea. One is that synthetic judgements predicate concepts of intuitions, and synthetic a priori judgements predicate pure concepts of pure intuitions. [See Allison, p.73; and Butts, 1981, p.139-40.] The idea would seem to be that application to experience is achieved by virtue of the forms of intuition being also forms of sensation; otherwise the application problem has not even been raised, let alone resolved. Thus transcendental schemata, qua pure intuitions, are the referents of schematised rather than pure concepts: schemata become "referents", not intuitions. Such a move seems possible if and only if schemata are indeed pure intuitions.

As I indicated above, Gram, at least, denies that this is compatible with the "third thing" approach. This involves a refusal to accept the coherence of the idea that anything can be both intellectual and sensible, universal and particular, and relies on a failure to distinguish two separate senses of "pure intuition" in Kant. These two senses—viz. "mere spatiality", or Kant's form of intuition; in contrast to "determinate space", or space as formal intuition—allow us to say that while both of these are pure intuitions, only "indeterminate" space (and time) is "purely sensible", as Gram's thesis requires. As we have seen, the notion of determinate intuition is central to Kant's thought; at least, the idea of determinate space is central to Kant's philosophy of mathematics. It is this determinate intuition which is of interest to the mathematician in the construction of concepts. It seems plausible to say that the pure (formal) intuition that is produced by the mathematician's constructive activity is both intellectual and sensible, universal and particular. In Kant's words, it is the "sensible presentation of a concept".

It is tempting—after all of this—to sympathise with Pippin, who remarks somewhat wearily that "...if there ever was a case of *obscurus per obscurius*, the attempt to explain schemata by means of pure intuitions would be it". [Pippin, p.142.] Nonetheless, I am not inclined to be sceptical about the schematism problem as a whole. Pippin often seems to capture the essence of Kant's convoluted remarks in the schematism chapter, yet is quite unable to find any independent justification for the argument. According to him, the Transcendental Deduction's results simply do not leave any "application problem"—if it works at all. Yet the crucial difference between the deduction and the schematism lies in the difference between what is generally true of objects, (objects *Überhaupt*, as we might say); and what is true of objects in a specific sense. The problem enters with respect to the discernment of specific features of our form of experience. Pippin suggests that the schematism is an attempt to explain how the understanding accomplishes what the deduction proved it must, and in doing so makes clear that for Kant the central problem of the schematism is a problem of judgement—specifically, the problem of judgemental application. Thus schemata provide the "truth content" for categories. According to this, the question of the schematism chapter with respect to "subsumption" must be how a sensible manifold can be conceptually determined; it is not a question of how instances are recognised as having some common predicate. Indeed, Kant thinks that the latter presupposes the former. In other words, before we can recognise that a number of instances have a common predicate, we must already have applied a concept to an intuition via the schematism, presumably because the former consists of abstracting from the manifold by means of judgements; this could not be achieved unless the manifold was itself subject to conceptualisation. So the schematism explains how the categories can have "sensible" meaning or significance. It is Kant's attempt to show how it could be that the pure concepts have *only* a sensible meaning, given that they cannot themselves be derived from experience.

The idea that the schematism is a "bridge" between the intellectual and the sensible is, I believe, a coherent one, and that in spite of attempts to collapse the problem into the conclusions of the Transcendental Deduction on the one hand, or to incorporate it into Kant's theory of pure intuition on the other, the so-called application problem will not so easily recede. In fact, that problem seems to have been shifted, not resolved, in the arguments just rehearsed.

The second of the results of the schematism chapter—that schemata are rules for construction, not images—must now be emphasised. It is well known that Kant frequently characterises concepts as rules, so the use of the same idea for schemata draws attention at once to their relationship with both "synthesis" and "imagination". And to insist that schemata are not themselves images prevents, or is intended to prevent, those problems that, notoriously, occupied Berkeley. An *image* would be determinately one kind of thing rather than another; in Berkeley's (and Kant's) example of a triangle, the presented figure would be either scalene, obtuse angled, etc. And there is no satisfactory solution to the problem of "abstract general ideas" so long as the object of attention is the empirical triangle.

But what is different in the case of a mathematical construction qua example is that the object of the geometer's attention is the schema, not the image. As I indicated above, through such a mathematical construction we effect an isomorphism between the a priori truths which belong to the concept "triangle", and the identifiable a priori conditions exemplified in the construction. It is intriguing that this idea finds application in another area of Kant's philosophy which itself relies upon something made, something produced—viz., his philosophy of art. It has been argued that this concept of construction, emphasising "exemplars" in art, provides a link between art and mathematics. [Crawford, in Cohen and Guyer, pp.151-78.]

What is useful in this context is Crawford's insistence on the schema as the object of attention. This might seem at first sight to be an obscure notion. In fact, it seems to me, on the contrary, to be one of the most familiar of experiences, especially in the context of learning. There is an attractive suggestion that, for Kant, learning a concept just is learning the rules of construction: to construct a concept is to learn by examples. In any pedagogical situation, we do not accept that a concept—say, "triangle"—has been learned unless what is understood is something beyond that which is given in any particular empirical figure. What is thereby understood is not, as a matter of fact, and cannot be, as a matter of logic, anything restricted to that empirical figure. To say that we have learned about the concept "triangle" is to say that our attention is drawn to those features of it not limited to any particular figure: at the same time, we have not learned the concept if all that we are able to say of it is confined to an abstract or "merely conceptual" knowledge of it. And there is indeed a creative—and often neglected—aspect of proving theorems in geometry which consists in having, so to speak, an "eye" for which axiom would be most appropriate to use as the foundation of a proof. [See also Crawford, *ibid.*] *Following* a proof—when one is merely passively responding to the steps used by a teacher—is not the same as *understanding* a proof; the latter requires that one construct something for oneself. In other words, one must know the concept in a purely logical sense and then apply it in a particular instance. And to employ Kantian terminology, in such circumstances the intuition would be "adequate to the concept".

I will return to this idea that schemata are rules for the construction of images in the special context of constructivism in Kant's philosophy of mathematics, and the argument from incongruent counterparts, below. For the moment, in order to harden the link between the latter and the schematism in general, I want to draw attention to some interesting parallels. Consider this list:

1) Schemata are rules for construction.

1)* Incongruent Counterparts can be actual constructions.

CHAPTER THREE – ACTS, INTUITIONS, AND CONSTRUCTIONS

2) Schemata are not themselves images.
2)* Incongruent Counterparts are not reducible to "intuitions".
3) Schemata particularise general concepts.
3)* Incongruent Counterparts are "representations" of concepts.
4) Schemata link concepts and intuitions.
4)* Incongruent Counterparts exist on the "interface" of concepts and intuitions.

(The starred propositions will be discussed in more detail below.)

I must point out here that I am not claiming that incongruent counterparts have these properties uniquely: for instance, Kant argues that there exist incongruent counterparts—hands, for example—that are not "constructions" in any sense I am concerned with here. It is nonetheless true that incongruent counterparts qua mathematical objects can be constructed. I will try to show how the ideas of construction and schematism assist in our understanding those objects qua mathematical objects, and claim no more for my interpretation than this.

The third of the results of the schematism that I suggested above are of permanent interest is the idea that schemata provide an essential connection with time. In some respects this may seem a consideration that it somewhat out of phase with the rest of the discussion, but it is in fact an essential element in Kant's methodological finitism. Since there is also an obvious sense in which Kant's idea that time is a pure form of intuition links with intuitionist philosophy of mathematics, (with Brouwer in particular), we need to examine this idea more closely.

To provide a bridge between sensibility and understanding we have seen that it is necessary to identify some third thing which is of sufficient generality to include the possibility of constructing concepts in pure intuition and the latter's forms, space and time. It is clear that time must be in some sense more fundamental than space: time is the "immediate" form of inner intuition, and also the "mediate" form of outer intuition. Put simply, the results of the Aesthetic imply that a merely temporal experience is possible, while the idea of a spatial but non-temporal experience must be rejected as incoherent. So to suggest a bridge between concepts and intuitions means that whatever is supposed to have this mediating function must be more than spatial. For Kant, the schemata must be time determinations. They are a universal feature of all appearances—"formally universal", and not just something empirical. And only time has this character of universal formality.

How are we to understand this, given what I have already emphasised, viz. that mathematical constructions, while not reducible to images, nonetheless seem to have more to do with spatial figures than with anything temporal? The important point is that the schema is a rule of construction—a rule, or set of rules, which permit the construction of an empirical figure. The schema as "object of attention" is not some ghostly mental image either, for even the mental image is subject to the same rules of construction as the empirical figure. When we understand the schema, the figure produced exemplifies the concept by means of it. Only in this way are we able to avoid the dilemmas of "abstract general ideas". And only in this way are we able to provide a coherent account of the constructions of mathematics that are not spatial in any obvious sense, viz., in arithmetic and algebra.

A modern analogy may help to remove the inevitable sense of artificiality from this thesis. We are (almost) all of us totally familiar with a set of rules—sequences—which issue in spatial analogues. Consider the relationship between a computer keyboard and a monitor. In principle, the state of the computer at any one time "contains" what it is that it "knows": it contains information in the form of a language which, were we clever enough, would be readable as information. The spatial representation (on the monitor) is "merely" a useful exemplification, in a different form, of that same information. But what the computer "knows", it knows independently of this representation. The analogy is even more precise (and interesting) in the case of computer graphics: these are clearly "spatial" representations, yet their realisation on the screen is as a result of sequences of instructions which are "spatial" in no philosophically interesting sense.

CHAPTER THREE – ACTS, INTUITIONS, AND CONSTRUCTIONS

So far as Kant's philosophy of mathematics is concerned, it is the *act of construction* that is of primary interest to us. In just the same way that Butts has suggested that Kant's use of examples throughout the critical philosophy has been too much neglected by commentators, I am suggesting that Kant's use of "act" terminology has been almost equally neglected.

We should note that Kant does not use the same word in all contexts, as is suggested by English translations. Sometimes the preferred term is *"Handlung"*, but perhaps the most interesting use is of the word *"Aktus"*. This emphasises a certain legislative power of the mind and is analogous with "acts" in the law. Legal metaphors abound in Kant, of course; the critical philosophy gives us "Deduction", *"quid juris"*, *"quid facti"* etc. But Kant's use of the concept of a legal act as one example of a mental act does not appear to have been much remarked upon. An exception is Smyth, who points out that "...the central question for criticism is always whether or not an act in the legal sense can be said to have been properly executed. Kant's term in the transcendental analytic for the fundamental act of the understanding, *Aktus*, is not the ordinary term for a mental act; he has borrowed the term for an official or legislative act". [Smyth, p.159.] To pursue the analogy further, for a construction to be "adequate to the concept", the "act" that produced it must accord with the conditions of the concept, otherwise the mathematical object will lack a certain "right": it may be logically legitimate, but it is only through an act—the schematism—that the concept can be afforded a legitimate use. We will find this "act" terminology making another appearance when we look again at incongruent counterparts.

My final point in relation to the schematism is merely to reiterate something raised briefly above, viz. the idea that through schemata, concepts and intuitions are linked in constructions, and that the a priori conditions exemplified in the construction are isomorphic with those a priori truths which belong to the concept thereby constructed. Should there be any looseness of fit between the concept and the intuitive construction, the intuition would no longer be adequate to the concept.

Without the intuitive construction, mere logical possibility would be sufficient to guarantee the legitimacy of a mathematical object: the problem of application would be not so much solved, as dissolved.

3.3. Kant's Constructivism

I first wish to identify six points which are related to what has already been said in relation to the schematism, but which can profitably be treated as independent positions for Kant's philosophy of mathematics. All I will claim for them is that they are to be found in Kant, and that taken together and in concert with other issues raised throughout this discussion, they provide a consistent model for thinking about Kant's views on space, time, and mathematics; as well as for a full understanding of the crucial distinction between reasoning from concepts (philosophy), and reasoning from the construction of concepts (mathematics).

These six theses are as follows; a) in mathematics, logic is not enough; b) running through Kant's thinking on mathematics is a distinction between real, and merely logical, possibility; c) mathematical objects are particulars which represent concepts; d) construction makes possible objects mathematically "real"; e) construction of mathematical objects is in pure intuition; f) certain mathematical symbols can be regarded as "sensuous epistemological tools".

It should by now be clear enough from what has already been discussed concerning the schematism, that for Kant, logic alone will not provide the materials for producing what we have in mathematics. Kant was firmly "anti-logicist"; he set himself against any philosophy that tried to reduce mathematics to logic. Mathematics has an autonomous subject-matter; it is *about* something, viz. certain structures and relationships which are only possible given the separation of concepts from intuitions, and the further possibility of constructing certain objects in space and time which are intuitive embodiments of conceptual truths. Kant's position that the transition from concepts to intuitions, (via the schematism), undercuts any reduction of the latter to the former. In the

CHAPTER THREE – ACTS, INTUITIONS, AND CONSTRUCTIONS

case of mathematical constructions, examples are employed as representations of concepts whose meaning is given in and through the mathematical system at hand.

There is a direct sense in which the mathematical judgements of a system are about the exemplars which the system itself makes constructively possible. [See Butts, p.269.] Butts thinks that is follows from this that the only constraints on construction are logical. But as I have already pointed out, Kant's philosophy not only permits, but actually demands the possibility of alternative logical systems and alternative geometries: this, as we have seen, is a consequence of the synthetic nature of mathematical propositions. The constraint on what can actually be constructed is not logical, therefore, but—if the expression can be allowed—"structural". That is, it is the limitations placed on the forms of our spatial and temporal intuitions that delimit which mathematically (logically) consistent concepts may actually be constructed. Put another way, it is the nature of these forms of intuition that constrains what is to be possible in these "first level" constructions. Obviously, we could not construct any concept which was internally contradictory; but this is not the same as saying that logic is itself a constraint on construction: logic constrains the internal meaning of the concept, but only construction in intuition is able to constrain the real meaning of the object. Thus constructions depend upon extra-logical considerations. This follows from the fact that Kant says that mathematical judgements are synthetic. And it is precisely to overcome the limits of "analytic" systems that Kant wants the objects of mathematics to be constructed. [Cf. Butts, p.273.]

In broad terms, therefore, we may affirm the second thesis above— viz. the distinction in Kant's philosophy of mathematics between what is merely logically possible, and what is real. Now since some such distinction is presupposed in the schematism—permitting as it does the move from the possession of a concept in its generality, to the application of that concept in a determinate intuition—and also in the doctrine of construction, it will clearly be required when the limits of constructibility are the issue in Kant's discussion of indirect proof. All that needs to be said at this stage is that some such distinction is available to Kant.

The third of the points to be reiterated is Kant's thesis—clear by now from the schematism—that mathematical objects are particulars which serve as representations of concepts. We do not require further arguments to those given in Chapter Two: I merely wish to draw attention to it again in this context; its relevance to the issues raised here should be obvious.

The next point to emphasise is Kant's thesis that construction makes a possible object "real". Butts makes the interesting suggestion that concepts are learned from constructed examples by our attending to the act of construction—by which Butts understands not some mental event, but the transaction involved in producing the example. There is clearly a resemblance between this and my earlier emphasis on "acts": yet the two views are not quite parallel. Is it clear that I learn the concept by constructing the example? I would prefer to say that I learn the significance or application of the concept by constructing the example. On my view, it is necessary to have the concept, and then try to apply it: this works, it must be emphasised, *only* in the case of mathematical concepts, since concepts or what Kant sometimes calls "definitions" may, in mathematics, precede their "objectification" in and through a construction. (This follows from the possibility of arbitrarily inventing a concept in mathematics.) As I see it, the construction would not be deemed adequate to the concept if one learned it in the construction. I take it that what Kant is saying amounts to a *post facto* explanation of what must occur when we do what we can already do, so to speak—a point that will be further supported below in my final discussion of incongruent counterparts.

The penultimate consideration listed above is that construction takes place in pure intuition. No further elaboration of this is required, since we have seen that it is in effect built into Kant's thesis that mathematical propositions are synthetic and a priori.

CHAPTER THREE – ACTS, INTUITIONS, AND CONSTRUCTIONS

Although pure concepts are known a priori, we have seen that mathematics must "hasten to intuition" in order to construct its objects. A construction in empirical intuition would not yield those characteristics Kant insists are bound up with our recognition of mathematical truths, viz. universality and a priori necessity. One complicating factor here is that Kant, notoriously, writes of space and time both as pure intuitions *simpliciter*, and as forms of pure intuition. The essence of the idea of mathematical construction consists in the affirmation that space as form of intuition is logically prior to space as "formal" intuition: mathematical constructions are in—one is tempted to say are "responsible for"— determinate space. The mathematician, in constructing a concept, appeals to this determinate space, rather than to space as "amorphous continuum".

It would therefore seem that when Kant insists that constructions are in pure intuition this could mean either that the concept, in being constructed, transforms space as form of intuition into space as formal intuition: or the process of construction could be taken as being in an already determinate space. The former seems more plausible, since such transformation of the indeterminate to the determinate in the special case of mathematics, is an example of the general function of the schematism in providing an image for a concept, and thus in moving from the nature of objects in general, to the nature of specific objects in a determinate intuition. And of course, this "objectifying" function of the schematism is precisely what I have been describing in the particular case of mathematical construction.

We should note that modern constructivist mathematics, in eschewing Kant's spatial intuition, leans more heavily on temporal intuition; but there is no obvious reason for thinking that Kant's constructivism is concerned only with the former. In suggesting that mathematical signs are "sensuous epistemological tools", Kant provides the material for a transition from construction in spatial intuition, to construction in temporal intuition. [P.C., p.24.] Such an idea would certainly permit Kant to regard constructions as, in essence, the intuitive embodiment of structural relationships. These may be spatial relationships, or they may be expressed for heuristic convenience in spatial constructions; but so long as the symbols express something general through what is particular, they can legitimately be regarded as constructions in pure intuition. Thus algebraic symbolism is constructive.

The problem with this is that if we accept the claim that an example, in Kant's terms, must be "more than a variable and less than a constant", it is harder to accommodate algebraic relationships within a general theory of constructions qua examples. It might even be thought *prime facie* plausible to take Kant's notion of mathematical signs as sensuous epistemological tools in a formalist, rather than constructivist mode. For instance, in that passage where the comparison is made, Kant says that the inferences and proofs of mathematics take place "concretely, under signs". He continues:

> For since mathematical signs are sensuous epistemological tools, it is possible to know that no concept has been neglected and that each single comparison has occurred, according to easy rules etc.—it is possible to know this with the same confidence with which one is assured of what one sees with one's eyes. The attention...does not have to consider things in their universal representation, but only the signs known individually and sensibly. [P.C. p.24.]

On the face of it, Kant seems to be arguing that the "transparency" of mathematical inference and proof—in contrast to the opaqueness of philosophical analysis of concepts—is due to the signs themselves being present to the senses, where what one sees—marks, symbols, etc.—is, so to speak, all that one has. As Kant puts it, the attention is not directed to "universal representation" but to signs known individually and sensibly. If one attaches this notion to the idea discussed above that the construction itself is an intuitive embodiment of a concept, one moves from Kant's "formalism" to a clear statement of constructivism.

This view of mathematical signs as sensuous epistemological tools has some links with both constructivism and formalism: it relates in an obvious way to the former, by means of the idea of intuition; it relates (not quite so happily) with the latter, since if one simply "unhooks" intuition from "sensuous", one is left with a bare sign, to be manipulated in purely formal arrangements in accordance with mathematical rules.

It has to be conceded that some of these issues will remain either difficult or obscure in the absence of a comprehensive Kantian account of the nature of symbols. The debate here makes a small contribution to such an account, in those discussions of sign, symbol, and schema; but Kant's language in this context makes any single interpretation elusive.

3.4 Incongruity and Constructions

I want now to discuss in turn a number of propositions that I hope will illuminate this perennially interesting and controversial topic. Although some of them were raised in the earlier discussion, I will rehearse the salient points again in as much as they contribute to the overview of Kant's philosophy of mathematics presented in this chapter. Any repetitions will, I hope, be justified by the context.

The permanently interesting results of Kant's various discussions of incongruent counterparts amount to the following: a) mathematical objects can have either real or merely logical existence; b) to be real, a mathematical object must be constructed in pure intuition; c) opposite-handed mathematical objects have identical "conceptual" descriptions; d) handedness is not a conceptual property; e) constructions are not reducible to "instructions"; f) distinguishing opposite hands requires an "act" in pure intuition; g) mathematical objects are neither merely intuitive nor merely conceptual; h) incongruent counterparts exist on the "interface" of concepts and intuitions; i) the construction of such objects involves elements of the schematism as outlined above.

I believe that put into this form these propositions capture the most important features of Kant's arguments; and I also believe that at the close of this discussion I will have offered good grounds for thinking that the argument is focally important for understanding Kant's constructivism. The detailed argument from incongruent counterparts, as discussed in 2.7 above, will be presupposed in what follows.

We can deal with a) and b) very briefly. For Kant, logical existence means freedom from contradiction. This being understood, all kinds of mathematical objects are possible—e.g. hypercubes, n-dimensional spaces, (where $n > 3$), non-Euclidean triangles etc. This criterion yields purely "formal" objects, where formal means logical, rather than related to intuition and its pure forms. The distinction between real and logical existence is analogous to the doctrine that construction transforms logical into real existence. In order to discover whether a syntactically consistent, logically non-contradictory mathematical object can be transformed into a "semantically" significant objected, we construct that object in pure intuition.

And thus we arrive at the second proposition above, viz. that mathematical reality consists in constructibility. (A question of logical priority would seem to be raised here: it would certainly look as if the thesis is that, to be significant just is to be capable of construction in pure intuition; yet it would also seem that we know that it is significant if and only if it is constructed.) We are reminded of Kant's insistence that "existence" is not a real predicate. Now although this was offered in the First Critique as a refutation of the ontological argument for the existence of God, the subsequent debate has both narrowed and broadened Kant's original thesis. Most recently, the question has been predominantly concerned with matters of formal logic.

We should recall Kant's thesis in any event. He writes that he is concerned to end "idle and fruitless disputations" concerning ontological proofs for God's existence. This "proof" is founded on an illusion caused by confusing *logical* with *real* predicates. He goes on:

> Anything we please can be made to serve as a logical predicate; the subject can even be predicated of itself; for logic abstracts from all content. But a determining predicate is a predicate which is added to the concept of the subject and enlarges it. Consequently, it must not be already contained in the

concept...'Being' is obviously not a real predicate; that is, it is not a concept of something which could be added to the concept of a thing. It is merely the positing of a thing, or of certain determinations, as existing in themselves...By whatever and by however many predicates we may think a thing, we do not make the least addition to the thing when we further declare that this thing *is*. [A597-A600/B625-628.]

Even those philosophers who have denied Kant's claim that existence is not a real predicate often concede that, nevertheless, it is a rather unusual one from the logical point of view. Rescher, for example, while affirming that "exists" is a predicate, links this claim with a general theory of possibility. The predicate "exists" is, in these terms, "inherently real-world bound". [Rescher, 1975, p.131.] In other words, this predicate cannot be transposed from one world to another. An existence proposition, if true, is true because of something obtaining in the actual world: "existence" simply represents that predicate which can truly be predicated only of members of the actual world. As Rescher says, it applies "universally and trivially to *all* members of this domain". [Rescher, *ibid.* p.129.] So if we do treat existence as a predicate, we find that it is one which is applicable only to items in the actual world. It would seem to follow from this that such propositions must have non-contradictory negations and are thus "synthetic" in Kant's sense. Retaining the idea of existence as a predicate may have, as Rescher indicates, certain advantages for logical formalisations.

The third proposition from the list above requires more elaborate consideration. Most discussions of incongruent counterparts concede Kant's point that two opposite-handed triangles have identical "conceptual" descriptions. [E.g. Buroker, Chapter 3.] Kant's point would therefore be that in such a case these objects have been transformed from logical to real existence; and that in so far as this has been achieved, some important property appears somehow to have been "discovered" on what might be called the "real" level, which could not have been present, or foreseen, at the conceptual or logical level alone, which nonetheless was

THE IDEAL AND THE REAL

sufficient to allow their purely logical existence. Kant's question would then be this; if this property—the property of handedness—did not come from logic, where did it come from?

The answer, of course, is that the property comes from the fact of its construction in pure intuition. Now although I will in fact agree with Kant that the two opposite-handed triangles do indeed have identical conceptual descriptions, it has been argued by Peter Alexander that Kant need not be conceded this crucial point at all. It is therefore important that this criticism be confronted, before we can feel entirely comfortable in proceeding on the basis that Kant is correct about this central thesis. [See P.Alexander, 1984-85.]

Alexander asks us to consider the following figure:

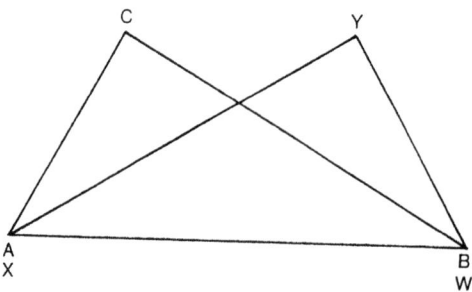

He next asks us to consider instructions for *constructing* this figure, as follows:

1) Draw straight line *AB* (=*WX*).
2) From *A* draw a line at 60° to *AB*.
3) From *B* draw a line at 30° to *AB* on the same side of *AB* until it intersects the line from *A* at *C*.

ABC is complete.

4) From *X* (*A*) draw a line at 30° to *XW* (*AB*) on the same side of *XW* as *C*.

166

CHAPTER THREE – ACTS, INTUITIONS, AND CONSTRUCTIONS

5) From *W (B)* draw a line at 60° on the same side of *XW* until it intersects the line from *X (A)* at *Y*.

WXY is complete.

Alexander concludes from this that in this form the instructions for triangles *ABC*, and *WXY* are different, "...so the two original triangles have different geometrical descriptions". [P.Alexander, p.9.] Now it should be perfectly clear from the discussion so far that I am anxious to affirm the importance of construction for the incongruent counterparts argument. Alexander's interpretation aims to show that a) using Cartesian axes two triangles which are incongruent counterparts *can* have identical descriptions; and/or b) that *constructing* incongruent counterparts gives them different "descriptions". Contrary to this, I want to insist that a) imports just the kind of conventional frame of reference that Kant believes *cannot* be part of the respective triangles' conceptual description; and that descriptions ≠ "constructions" for these purposes, even if a) was correct.

From this example, it does at first sight appear that we have been provided with identical descriptions for opposite-handed triangles. But this works only because we already have in front of us two *different* triangles. Alexander is therefore *reconstructing* these triangles, and probably begging the question against Kant by introducing from the outset a conventional way of distinguishing one triangle from the other, (*"ABC"* etc.) The process of construction, (given the plausible, but not entirely comfortable identification of "instructions" and "constructions"), does not determine what kind of triangle—left- or right-handed—will be produced. Everything depends upon the point at which the convention is introduced. The instructions may, but need not, produce incongruent counterparts. It does not follow from this that incongruent counterparts have identical *descriptions,* but that they may be constructed employing identical *instructions.* It further follows from this that constructions qua instructions are not equivalent to descriptions: which is exactly Kant's point.

Alexander's argument can therefore be seen as a version of Kant's own thought-experiment with the single hand alone in the universe discussed in 2.7, above, which I will rehearse here for convenience. Imagine, writes Kant, that the universe is empty except for a single hand. Is this a left-hand, or a right-hand? Given the conceptual indeterminacy of handedness, it does not seem quite right to assert one or the other. But now imagine that a handless body is created: it seems that the hand must fit on either the left or the right wrist of the body. Yet if it fits, say, on the left wrist, then Kant argues that it must have been a left hand *before* the introduction of the handless body. The determinate handedness of the first object, is given retrospectively after the second object—which provides the convention—is introduced. I believe that Alexander's analysis would yield the results he wants only if the instructions unambiguously produced the same results. However, the instructions do not contain enough information to specify the handedness of any triangle: so the fact that they might be differently handed cannot be explained by the instructions alone. Only an arbitrary or conventional decision can do that, as Kant recognised.

From this, I can now reaffirm my next point, which is that handedness is not a conceptual property: it is introduced at what I have called, crudely, the "semantic" level. What this means is simply that the significance of incongruent counterparts is irreducibly connected to what is, in Kant's terms, phenomenal or intuitive. Can we then conclude from this that handedness has something to do with the act of construction as such? When faced with the challenge, "Construct a triangle", the logical/conceptual constraints do not include the object's being handed one way or the other. (It does not even include Nerlich's concept of being an "enantiomorph" as such, since not all triangles are handed.) We can still construct the triangle in accordance with the logic for "possible triangles". (It is as if handedness is a kind of "floating" parameter.) Since the logic for "possible triangle" cannot include an instruction to the effect that the object, when constructed, must be handed one way or the other, or not handed at all, we introduce the hypothesis that it is Kant's "pure

CHAPTER THREE – ACTS, INTUITIONS, AND CONSTRUCTIONS

intuition" that permits the additional property to be introduced at the "constructive" level. Following from this, we can say that what is true of the construction cannot be reduced to mere concepts. Constructions are not reducible—without loss of information—to "instructions", *pace* Alexander.

The next point reintroduces the "act" locutions noted above. What Kant says is that distinguishing opposite hands requires an act in pure intuition. Certainly some kind of act is required to construct the triangle: does this mean that distinguishing them also requires an act of an analogous kind? And why an act in pure intuition? We can see directly that in Kant's terms this act could not be in empirical intuition, because the objects constructed in mathematics have properties not available in empirical intuitions alone were involved—viz. universality, and necessity. So the thesis that the act is in pure intuition applies both to constructing and "noticing". Just as the logic leaves out something important in the instructions for producing the triangles, so our perception of them—that is, our perceptual distinguishing of them—simply "reads off" what the construction now contains, as well as what is true of the triangles from a merely conceptual point of view. Yet in spite of this (apparent) placing of handedness in the mind of the perceiver, the triangles are perceived as having some *internal* property which distinguishes them. It is the triangles that are different, not simply our perceptions of them.

For Kant, mathematical objects are neither merely intuitive nor merely conceptual. It is clear from this discussion that such objects are not constituted by purely conceptual characteristics; but what prevents them from being taken as wholly intuitive? Why not think of such objects as having their meaning exhausted by intuitive considerations? Would this not provide the kind of link that Brouwer affirms between his constructivism, and Kant's? What appears to prevent this possibility is that consideration discussed earlier to the effect that mathematical objects are given logical possibility as well as the possibility of construction. In this way, Kant may consistently be taken as having included (or at least, not excluded) the possibility of non-Euclidean geometries etc., where the intuition in question is not ours, and where certain kinds of constructions are impossible.[1] On a "global" scale, any mathematics will be an amalgam of the logical and the intuitive; the type of spatial and temporal intuition possessed by any finite minds will determine what can be constructed, and hence will determine what will have real, in contrast to merely logical, existence. Of course, if we have no interest in accommodating non-Euclidean geometries etc., then mathematical objects could, in principle, be exhausted by an intuitive constructivism of a Brouwerian kind. Operations on such objects would still need to be consistent, however, so the question would then be whether intuition alone could guarantee this consistency. It might be argued that the idea of the possibility of construction in pure intuition would itself be some guarantee of consistency. But since it is Kant's version of constructivism rather than Brouwer's with which I am concerned here, this issue will not be pursued any further.

My final point in relation to the problem of incongruent counterparts was expressed earlier in the highly metaphorical idea that such objects exist at the "interface" of concepts and intuitions. What this idea is intended to convey is the sense that incongruent counterparts seem to possess what might be thought of as "transitional" properties. They are not wholly reducible to logical/conceptual characteristics; yet in themselves they do not permit any denial that only intuition was necessary.

The point made above was that there are indeed instructions for producing such objects; but that such instructions must either specify handedness—which is not itself one of the object's logical properties—or, so to speak, leave it to the "constructor"; and then no rule could determine whether the object shall be left- or right-handed. To push the metaphor further—that these objects are "transitional"—we could summarise the idea by contrasting a left-handed *triangle*, with a *left-handed*

---

[1] Stephan Körner, in his (1984), p.74, insists that "Kantians" must reject non-Euclidean geometries for metaphysical reasons, while accepting elsewhere that Kant could consistently hold that such geometries are logically possible. See Körner, 1968, p.139.

triangle. The former places the descriptive bias onto the conceptual characteristics, while the latter reverses the bias. It seems that mathematically the latter presupposes the former, which is also the order of priority for Kant in his move from concepts to intuitive constructions, via schemata. Again, the conclusion to which we are impelled is that handedness is the outcome of some kind of act of choice, constrained by intuition, and performed on an object which nonetheless seem actually to possess this property in a full-bodied sense.

3.5 Indirect Proof

Given the close relationship in modern constructivist philosophy of mathematics between the general idea of a proof, and the refusal to countenance certain "non-constructible" objects involving actual infinities, it is to be expected that Kant's constructivism entails similar constraints on the range of legitimacy of mathematical objects. Where Kant differs is in not discussing the two issues in the same context. It would therefore be quite easy to underestimate the extent to which Kant's positive characterisation of the nature of mathematical reasoning entails negative restraints with regard to Ideas of Reason and infinite totalities. As a conclusion to this discussion I want to emphasise this "negative" aspect of Kant's constructivism which, though placed by him within more general concerns of transcendental philosophy, has clear implications for the special case of mathematics. The proposition to be thus emphasised is simply to say that for Kant, in the absence of the possibility of a constructive proof, or of construction of certain objects *per se*, only indirect of "apogogic" proof can be employed; and that the use by Kant of apogogic proofs allows the immediate inference that no constructive proof is possible at all for the objects in question.

Kant's primary employment of apogogic proof is in the Antinomies of Pure Reason: these deal in one way or another with the concept of infinity. We can say that Kant is consistent with his own principles in not even discussing proofs that he wishes to repudiate in anything other than an indirect mode. Indeed, if Kant had been able to argue from some direct method in this respect, he would *ipso facto* have jettisoned his own conviction that these were indeed *antinomies*. I will close this section with a speculation on the nature of antinomies in relation to Kant's repudiation of so-called "subreptic" fallacies.

To say that no constructive proof is possible when we are dealing with infinities is a clear consequence of the previous discussion. This is the case whether the infinity involved is mathematical or "philosophical", as in the Antinomies. The absence of any of the conditions and constraints placed on construction as discussed above entails that in so far as a proof is possible at all, that proof must be indirect. In fact, Kant's position is quite straightforward, at least in principle. Construction is what turns logical into real existence; real existence is constrained by pure intuition. If pure intuition is somehow or other excluded from the outset, then only mere logical possibility is available. In the example of the Antinomies, pure intuition is excluded because the Categories are not being legitimately applied to intuitions; indeed, because infinities are involved, no such application is possible at all. What this amounts to is that no schema could mediate between the concept and the intuition, hence no determinate object can be given to experience. If no determinate object can be given, no real mathematical object can be given either, because real mathematical objects are always determinate being, in Kant's sense, particulars constructed in intuition. It is thus intuition that provides the very possibility of constructive proof. Without intuition, some "proof" is nonetheless possible, viz. indirect or apogogic proof. Indirect proof is thus by definition non-constructive. Infinity is no more a concept that constructive mathematics can deal with, than it is an object that the categories can deal with in the Antinomies. Real knowledge arises when categories are applied, via schemata, to intuitions: real knowledge comes in mathematics when schemata mediate between the general concept of an object, (its logical possibility), and the intuition which embodies it.[2]

---

[2] For Brouwer, this intuition is temporal: by taking the schematism account more seriously, Brouwer's link with Kant seems to be even more striking.

CHAPTER THREE – ACTS, INTUITIONS, AND CONSTRUCTIONS

As Kant puts it, the apogogic method of proof is only permissible in those sciences where it is impossible to erroneously substitute the subjective for the objective. [A791/B819] It has been suggested that this amounts to a general restriction on the use of apogogic proof, such that it is an allowable form of argument only when there is no danger of confusing epistemological and ontological questions. [Srzednicki, p.76.] Although there is some justice in putting the issue like this, it seems more fruitful to think of Kant's assertions on this question as related to his frequent condemning of those writers guilty of what Kant terms the "subreptic fallacy".

There is an extended and, it must be admitted, not entirely clear discussion of "subreption" in the *Inaugural Dissertation*. What the subreptic fallacy amounts to is the illegitimate predication of a "sensible" concept to an intellectual concept. Kant warns that "...great care must be taken lest the domestic principles of sensitive cognition transgress their boundaries and affect things intellectual". [P.C., p.81.] We can see the connection between subreption, and the solution to the Antinomies, in the following passage, where Kant is discussing the "Second Class" of subreptic axioms:

Since every quantity and series whatsoever is only cognised distinctly through successive co-ordination, the intellectual concept of a quantity and a manifold arises only with the help of this concept of time, and it never reaches completion, unless the synthesis could be achieved in a finite time. Hence it is that an infinite series of co-ordinates could not be comprehended distinctly according to the limits of our intellect and so by the fallacy of subreption such a series would appear impossible. [P.C. p.86.]

We must not take these early ideas as in any sense definitive: what we can say is that the kinds of problems, later classified by Kant as antinomial conflicts, had already been perceived as the source of substantial metaphysical confusion—the confusion of subreption. One way of committing the subreptic fallacy would be to deny even logical existence to a concept—and thus to declare it to be "absolutely impossible"—as a result of a defective application of a sensible predicate to an intellectual concept: in other words, to conflate the ontological with the epistemological. Some such failure is evident throughout the Antinomies; it should not be forgotten that the arguments presented there are not regarded by Kant as valid due to their failure to distinguish appearances from things in themselves.

Whether or not a proof is legitimate will depend, for Kant, on whether the objects are constructible. In mathematics, where the idea of the merely logically possible is unhooked from the domain of the totally unknowable thing in itself, there will always be a legitimate role for indirect proof. In philosophy, which is reasoning from concepts and not from the construction of concepts, there is the perennial danger of subreption, where predications are made of the intellectual by the sensible. It is to avoid such fallacies that the focal conception of possible experience is required; and it is to the Transcendental Analytic that Kant asks us to look for alternative methods of proof.

Bibliography

In order to keep this listing to a minimum, I have included here only primary sources; and those secondary sources to which reference is made directly in the text, or to works found particularly useful in its preparation.

Primary Sources

Alexander, H.G., *The Leibniz-Clarke Correspondence*, Manchester: Manchester University Press, 1970.
Kant, Immanuel. *Logic*, translated by R.Hartmann and W. Schwarz, New York, 1974.
Metaphysical Foundations of Natural Science, Indiana: Bobbs-Merrill, 1970.
Prolegomena, Manchester: Manchester University Press, 1966.
Selected Pre-Critical Writings, Manchester: Manchester University Press, 1968.
Kritik der Reinen Vernunft, 2 vols. Frankfurt: Suhrkamp Verlag, 1974.
Critique of Pure Reason, translated by Norman Kemp-Smith, London: Macmillan, 1970.
Critique of Practical Reason, translated by T.K.Abbot, Longmans, 1898.
Critique of Judgement, translated by James Creed Meredith, Oxford: Clarendon Press, 1973.
Philosophical Correspondence, edited by A.Zweig, Chicago: University of Chicago Press, 1967.
Leibniz, G. von W. *Die philosophischen Schriften von G.W.Leibniz*, edited C.J.Gerhardt, Berlin: 1875-90.
Philosophical Writings, London: Everyman Edition, 1968.
The Monadology and Other Writings, trans. by R.Latta, Oxford: Clarendon Press, 1898.

Philosophical Papers and Letters, edited by L.Loemker, Reidel, 1956.

Newton, Isaac., *Mathematical Principles of Natural Philosophy*, translated by A.Motte [1729], revised by F.Cajori, University of California Press, 1960.

Secondary Sources

F.B. D'Agostino, "Leibniz on Compossibility and Relational Predicates", *Philosophical Quarterly*, 26, (1976).

S. Al-Azm, *Kant's Theory of Time*, New York: Philosophical Library, 1970.

Alexander, Peter. "Incongruent Counterparts and Absolute Space", *Proceedings of the Aristotelian Society*, LXXXV, 1984-85.

Allison, Henry. "Transcendental Schematism and the Problem of the Synthetic A Priori", *Dialectica*, 35, 1, 1981.

Kant's Transcendental Idealism, New Haven: Yale University Press, 1983.

Benacerraf P. and Putnam, H. (eds.) *Philosophy of Mathematics*, 2nd edition, Cambridge: Cambridge University Press, 1983.

Benardeté, José. *Infinity*, Oxford, 1964.

Bonola, R. *Non-Euclidean Geometry*, New York: Dover, 1955.

Brentano, Franz. *Philosophische Untersuchungen zu Raum, Zeit und Kontinuum*, edited by S.Körner and R.Chisholm, Hamburg: Felix Meiner, 1976.

Brittan, Gordon. *Kant's Theory of Science*, Princeton, N.J: Princeton University Press, 1978.

Broad, C.D. *Leibniz: An Introduction*, Cambridge: Cambridge University Press, 1975.

Kant: An Introduction, Cambridge: Cambridge University Press, 1978.

Buchdahl, Gerd. "Science and Logic: On Newton's Second Law of Motion", *British Journal for the Philosophy of Science*, 2, 1951.

Metaphysics and the Philosophy of Science, Oxford: Blackwell, 1969.

"Reduction-Realization: A Key to the Structure of Kant's Thought", *Philosophical Topics*, 12, 2, 1981.

Buroker, Jill Vance. *Space and Incongruence*, Holland: Reidel, 1981.

Butts, Ronald. "Rules, Examples and Constructions: Kant's Theory of Mathematics", *Synthese*, 47, 2, 1981.

Butts, R. and J. Davis. (eds.) *The Methodological Heritage of Newton*, Oxford: Blackwell,, 1970.

Cassirer, Ernst. *Leibniz' System in seinen wissenschaftlichen Grundlagen*, Marburg, 1902.

"Kant und de modernen Mathematik", *Kant-Studien*, 12, 1901.

Casteñeda, Carlos. "The Status of Arithmetical Propositions", *Philosophy and Phenomenological Research*, 21, 1960.

Cohen, Ted, and Paul Guyer. (eds.) *Essays in Kant's Aesthetics*, Chicago: University of Chicago Press, 1982.

Cohn, J. *Geschichte des Unendlichkeitsproblems*, Olms, 1960.

Courant, J. and H. Robbins, *What is Mathematics?*, Oxford, 1977.

Cox, C.B. "A Defence of Leibniz' Spatial Relativism", *Studies in History and Philosophy of Science*, 6, 1975.

Earman, John. "Kant, Incongruent Counterparts and the Nature of Space and Space-Time", *Ratio*, 13, 1971.

Erlichson, H. "The Leibniz-Clarke Controversy: Absolute versus Relative Space and Time", *American Journal of Physics*, 35, 2, 1967.

Folse, H.J. "Kantian Aspects of Complementarity", *Kant-Studien*, 1978.

Frankfurt, H. (ed.) *Leibniz: A Collection of Critical Essays*, Notre Dame, 1976.

Frege, G. *On the Foundations of Geometry and Formal Theories of Arithmetic*, translated by Eike-Henner W. Kluge, New Haven;Yale University Press, 1971.

Frey, Gerhard. *Erkenntnis der Wirklichkeit*, Stuttgart: Kohlhammer, 1965.

Friedman, L. "Kant's Theory of Time", *Review of Metaphysics*, VII, 1954.

Gent, W. "Leibnizens Philosophie der Zeit und des Raumes", *Kant-Studien*, 31, 1926.

Gil, D. "Intuitionism, Transformational Generative Grammar and Mental Acts", *Studies in History and Philosophy of Science*, 14, 3, 1983.

Hall, A.R. & M.B.Hall, *Unpublished Scientific Papers of Isaac Newton*, Cambridge: Cambridge University Press, 1962.

Hacking, Ian. "The Identity of Indiscernibles", *The Journal of Philosophy*, May 1975.

Hankins, T.L. "Algebra as Pure Time: William Rowan Hamilton and the Foundations of Algebra"; in P. Machamer and R. Turnbull (eds.) *Motion and Time; Space and Matter*, Columbus Ohio, 1976.

Hendry, J. "The Evolution of William Rowan Hamilton's View of Algebra as the Science of Pure Time", *Studies in History and Philosophy of Science*, 15, 1, 1984.

Heyting, A. "Constructivity in Mathematics", *Proceedings of Colloquium*, Amsterdam: North Holland, 1957.

Hintikka, Jaako. "Kant's 'New Method of Thought' and Theory of Mathematics", *Ajatus*, 27, 1965.

"Kant on the Mathematical Method", *The Monist*, 51, 1967.

"On Kant's Notion of Intuition", in *The First Critique*, ed. T.Penelhum and J.H.MacIntosh, California University Press, 1969.

Horstmann, R. "Space as Intuition and Geometry", *Ratio*, XVIII, 1976.

Humphrey, Ted. "The Historical and Conceptual Relations between Kant's Metaphysics of Space and Philosophy of Geometry", *Journal of the History of Philosophy*, 11, 1973.

Jammer, Max. *Concepts of Space*, Cambridge, Mass; Harvard University Press, 1954.

Kemp-Smith, Norman. *A Commentary on Kant's Critique of Pure Reason*, London; Chivers, 1969.

Kitcher, P. *The Nature of Mathematical Knowledge*, Oxford: Oxford University Press, 1984.

Körner, Stephan. *Kant*, London: Penguin Books, 1955.

"On the Kantian Foundations of Science and Mathematics", *Kant-Studien*, (1966).

Experience and Theory, London: Routledge, 1966.

The Philosophy of Mathematics, London: Hutchinson University Library, 1968.

"Introduction to Papers on Philosophy", in *Selected Papers of Abraham Robinson*, ed. S. Körner, New Haven: Yale University Press, 1978.

Koyré, Alexander. *From the Closed World to the Infinite Universe,* Johns Hopkins University, 1957.
Newtonian Studies, London, 1965.

Kripke, Saul. "Naming and Necessity", in *Semantics and Natural Language,* Synthese Library, 1972.

Lacey, H. "The Scientific Intelligibility of Absolute Space", *British Journal for the Philosophy of Science,* (1970).

Laymon, R. "Newton's Bucket Experiment", *Journal of the History of Philosophy,* (October 1978).

Lehman, H. *Introduction to the Philosophy of Mathematics,* Oxford: Blackwell, 1979.

Losee, John. *A Historical Introduction to the Philosophy of Science,* Oxford: Oxford University Press, 1977 etc.

Lucas, J.L. *A Treatise on Time and Space, London:* Methuen, 1973.

Martin, Gottfried. *Kant's Metaphysics and Theory of Science,* Manchester: Manchester University Press, 1955.
Leibniz: Logic and Metaphysics, Manchester: Manchester University Press, 1967.

McGuire, J.E. "'*Labyrinthus Continui*': Leibniz on Substance, Activity and Matter", in *Motion and Time, Space and Matter,* ed. P.Machamer and R.Turnbull, Columbus: Ohio State University Press, 1976.

MacLaurin, Colin. *An Account of Sir Isaac Newton's Philosophical Discoveries,* (4 volumes), London, 1748.

Meerbote, R. "Kant on Intuitivity", *Synthese,* 47, 2, (May 1981).

Nagel, G. *The Structure of Experience,* Chicago: University of Chicago Press, 1983.

Nerlich, Graham. *The Shape of Space,* Cambridge: Cambridge University Press, 1976.

Øhrstrøm, P. "W.R.Hamilton's View of Algebra as the Science of Pure Time and his revision of this view", *Historia Mathematica,* 12, 1, (1985).

Pap, Arthur, *The A Priori in Physical Theory,* New York, 1946.

Parkinson, G.H.R. *Logic and Reality in Leibniz's Metaphysics,* Oxford: Clarendon Press, 1965.

Pippin, R.B. *Kant's Theory of Form*, New Haven: Yale University Press, 1982.

Poincaré, H. *Science and Hypothesis*, New York: Dover, 1952.

Rescher, N. *The Philosophy of Leibniz*, New Jersey; Prentice-Hall, 1967.

A Theory of Possibility, Oxford: Blackwell, 1975.

Russell, Bertrand. *A Critical Exposition of the Philosophy of Leibniz*, Cambridge: Cambridge University Press, 1900.

Sklar, L. *Space, Time and Space-Time*, California, 1974.

"Absolute Space and the Metaphysics of Theories", *Nôus*, 6, (1972).

Smyth, R. *Forms of Intuition*, The Hague: Martinus Nijhoff, 1978.

Srzednicki, J.T. *The Place of Space and Other Themes*, The Hague: Martinus Nijhoff, 1983.

Stevenson, L. *The Metaphysics of Experience*, Oxford: Oxford University Press, 1982.

Sumner, L.W. and J.Woods, *Necessary Truth: A Book of Readings*, N.Y: Random House, 1969.

Tamny, Martin. "Newton, Creation, and Perception", *Isis*, 70, (1979).

Vasiliev, A.V. *Space, Time and Motion*, London, 1924.

Walker, R.C.S. *Kant*, London: Routledge, 1978.

Walsh, W.H. *Kant's Criticism of Metaphysics*, Edinburgh, 1975.

Warnock, G.J. "Conceptual Schematism", *Analysis*, 9, (1948-49).

Westfall, R. *Force in Newton's Physics*, Macdonald, 1971.

Weyl, H. *Philosophy of Mathematics and Natural Science*, Princeton: Princeton University Press, 1949.

Symmetry, Princeton: Princeton University Press, 1952.

Wilder, R.L. *Evolution of Mathematical Concepts*, Transworld Student Library, 1974.

Winterbourne, A.T. "Algebra and Pure Time: Hamilton's Affinity with Kant", *Historia Mathematica*, 9, (1982).

"Art and Mathematics in Kant's Critical Philosophy", *British Journal of Aesthetics*, 28, 3, (1988).

Wolff, R.P. *Kant's Theory of Mental Activity*, Gloucester, Mass: Peter Smith, 1973.

Index of Names

d'Abro, A., 7 n.3
Alexander, Peter, 127 n.18 166-169,
Allison, H., 60 n.5 149,

Benardeté, José, 131
Berkeley, George, 26, 87 n.11 153,
Broad, C.D., 14
Brouwer, L.E. J., 102, 103, 169, 170, 172 n.2
Buchdahl, Gerd, 28, 41 n.19, 77 n.9, 87 n.11
Butts, R., 159, 160

Cantor. Georg, 131
Cox, Chana, 39
Crawford, Donald, 82 n.10 154,

Einstein, A., 23
Earman, John, 115 n.17

Gram, Moltke, 151

Hacking, Ian, 16 n.8
Hamilton, W.R., 92 n.13
Hankins, Thomas, 92 n.13
Hanson, N.R., 61 n.6

Heidegger, M., 146
Hilbert, David, 102, 103, 104, 128, 134
Hintikka, Jaako, 82-84, 93
Huygens, Chr., 20

Ishiguro, Hidé, 36 n.15

Jammer, Max, 1, 7-10

Kemp Smith, Norman, 58 n.4, 100, 101
Kitcher, P., 65 n.8
Körner, Stephan, 97 n.15, 170 n.1
Koyré, Alexander, 22
Kripke, Saul, 57 n.2

Laymon, R., 7 n.4, 11 n.5
Locke, John, 28
Losee, John, 7 n.3

Mach, Ernst, 24
McGuire, J.E., 41 n.18
More, Henry, 2

Nerlich, Graham, 111, 112, 114, 115, 168

Øhrstrøm, Peter, 92 n.13

Pap, Arthur, 22 n.11
Pippin, Robert, 152
Poincaré, H., 98

Rescher, Nicholas, 32, 36, 40 n.17, 165
Russell, Bertrand, 40

Sklar, Lawrence, 9, 16
Smyth, R.A., 157
Srzednicki, J.T., 173

Tamny, Martin, 6

Walsh, W.H., 77
Westfall, Richard, 21, 27, 46 n.21
Weyl, Hermann, 115, 116
Wolff, R.P., 146

Zeno, 43, 46, 130, 136

www.ingramcontent.com/pod-product-compliance
Ingram Content Group UK Ltd.
Pitfield, Milton Keynes, MK11 3LW, UK
UKHW021302180426
11947UKWH00015B/977